Façades

Principles of Construction

The following titles have been published in this series:

Components and Connections in Architecture – Principles of Construction
Maarten Meijs, Ulrich Knaack, Tillmann Klein
ISBN 978-3-7643-8669-6

Prefabricated Systems – Principles of Construction
Ulrich Knaack, Sharon Chung-Klatte, Reinhard Hasselbach
ISBN 978-3-7643-8747-1

Façades

Principles of Construction

Ulrich Knaack, Tillmann Klein,
Marcel Bilow, Thomas Auer

Second and revised edition

We would like to thank Delft University of Technology for the financial support of this publication.
We would also like to thank Ria Stein for her editorial guidance as well as the students Jean-Paul Willemse, Vincent van Sabben,
Thijs Welman, Max Ernst and Farhan Alibux for their help in generating the drawings.

Layout and cover design: Oliver Kleinschmidt, Berlin
Translation into English: Usch Engelmann, Rotterdam
Subject editors for the first English edition: Thomas Schröpfer, Limin Hee, Singapore
Editor: Ria Stein, Berlin

Library of Congress Cataloging-in-Publication data
A CIP catalog record for this book has been applied for at the Library of Congress.

Bibliographic information published by the German National Library. The German National Library
lists this publication in the Deutsche Nationalbibliografie;
detailed bibliographic data is available in the Internet at <http://dnb.ddb.de>.

This book is also available as an E-Book (ISBN 978-3-03821-145-7).
It is also available in a German edition (ISBN 978-3-03821-094-8)
and a German E-Book (ISBN 978-3-03821-026-9).

First edition 2007
Second and revised edition 2014

© 2014 Birkhäuser Verlag GmbH, Basel
P.O. Box 44, 4009 Basel, Switzerland
Part of Walter de Gruyter GmbH, Berlin/Boston

Printed on acid-free paper produced from chlorine-free pulp. TCF ∞

Printed in Germany

ISBN 978-3-03821-044-3

9 8 7 6 5 4 3 2
www.birkhauser.com

CONTENTS

7 | 1 Introduction

14 | 2 From Wall to Façade

14 | Solid wall construction
16 | Warm façade, cold façade

16 | Openings in solid wall construction
18 | Bridging the gap
19 | Single glazing
20 | Box window
21 | Insulated glazing

22 | Walls with skeletal structure
22 | Half-timbered construction
23 | Platform and balloon framing

24 | Resolution of the wall into
     loadbearing structure and façade
25 | Post-and-beam façade
26 | Post façade
26 | Beam façade
27 | Curtain wall
28 | System façade

29 | Double façades
30 | Second-skin façade
30 | Box-window façade
31 | Corridor façade
32 | Shaft-box façade
33 | Alternating façade
34 | Integrated façade

36 | 3 Principles of Construction

37 | Areas of construction
38 | Façade bearing structures and load transfer
42 | Grid and positioning of the façade
     within the building
44 | Systems used in façade construction
45 | Post-and-beam construction
46 | Unit system façade
46 | Designing with systems

47 | Openings in façade constructions
47 | Hardware
48 | Windows

50 | Assembly

52 | 4 Detailing and Tolerances

54 | Building grid and positioning of components
56 | Combination of functions
57 | Detailing principles
57 | Layering of details

58 | Examples of detail development
59 | Masonry cladding
59 | Post-and-beam façade
60 | Unit system façade
61 | Parapet
62 | Plinth unit
63 | Joints

67 | Tolerances

70 | **5 Climate and Energy**

70 | **Façade as interface to the exterior**
70 | **Functional requirements**
71 | Thermal requirements
72 | Visual requirements
73 | Hygienic requirements
73 | Acoustic requirements

74 | **Regulating the comfort level**
     with the façade
74 | Ventilation
77 | Heating
78 | Cooling

80 | **Sun and glare protection**
84 | **Light-directing systems**

85 | **6 Adaptive Façades**

85 | **Sun**
86 | Light
86 | Heat
87 | Greenhouse effect

87 | **History of adaptive façades**
90 | **Collector façade**
90 | Trombe wall
91 | Transparent heat insulation
92 | Exhaust-air façade

93 | **Double façade**
94 | Box-window façade
95 | Shaft-box façade
96 | Corridor façade
98 | Second-skin façade

100 | **Alternating façade**
100 | **Integrated façade**

102 | **7 Case Studies**

102 | **Rear-ventilated façade**
      Concept House, RDM Campus, Rotterdam
106 | **Solid façade**
      State Archive Nordrhein-Westfalen, Duisburg
110 | **Post-and-beam façade**
      New building for the Department for Architecture
      and Interior Design at the University of Applied
      Sciences, Detmold
114 | **Unit system façade**
      Headquarters Süddeutscher Verlag, Munich

118 | **8 A Look Into the Future**

120 | **Material and construction**
124 | **Climate, comfort, energy**
126 | **Production and assembly**
127 | **(Design) tools**

Appendix
128 | **Authors**
129 | **Selected bibliography**
130 | **Index**
131 | **Illustration credits**

1 | Introduction

With *Façades – Principles of Construction* being the title of this book – now in its second edition –, one might begin to wonder why there is the need for yet another book about façades. As it is, there are sufficient volumes focusing on topics such as transparent façades, double façades or material-specific façade constructions (1, 2). The subtitle – *Principles of Construction* – should shed some light on the purpose of this particular book: it is designed for architects and students who wish to concentrate on the design principles of façades in a more fundamental manner. This book does not focus on specific façade variants; rather it explores basic façade systems, their origin as well as the principles of construction, building structural aspects and the integration with the building itself. The goal is neither a collection of design examples nor a compilation of current and regulation-conforming details, but to create a basic understanding of the façade and its technical realisation. Not based on specific European code norms and technical regulations nor dependant on specific material-related parameters, this understanding will enable the reader to analyse specific project examples with the aim to realise their own developments in a technically sound manner.

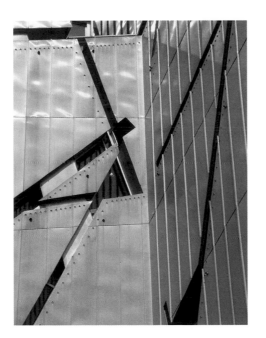 

1

2

Jewish Museum, Berlin,
Daniel Libeskind, 1999
Façade detail. The architectural concept envisioned a homogenous sheet-metal façade, which, during technical realisation, underwent a metamorphosis to a multi-layered rear-ventilated façade with embedded rain drainage and spillway.

Guggenheim Museum, Bilbao,
Frank O. Gehry, 1997
Geometrically complex building junction: solving the geometry and implementation of a façade system is part of the architect's responsibility but without modifying the structural system of the post-and-beam façade.

This approach is part of an integral design concept: an architectural design not only includes the concept, the composition of space and the organisation of the building's functions but should encompass its structural realisation. The definition of surface and structural materials and their detailed application is an expression of the building as a whole. Thus, the detail is part of the architectural concept, to be understood as an element on a special scale. The architect needs to exercise creative control of this element; otherwise, the detail will develop randomly and might influence the architectural expression contrary to its original conception.

Today the architect can no longer control every detail in its technical entirety – the range of technological developments and product diversity has become too broad. This book will provide an overview of typical solutions, the underlying systems as well as their functionalities. This information will allow the architect to be a competent partner in façade design. It will enable him or her to understand the suitability of each system in a specific part of the design and to determine its technical and geometrical limits.

We don't see the façade as an isolated building component but as an integral element with considerable importance in terms of the building's appearance. It should include additional functions such as loadbearing, active or passive environmental control (3) and individual creative expression (4).

3

Debitel Headquarters, Stuttgart,
RKW Architektur + Städtebau, 2002
Example of integrative planning of architecture and environmental concept: this alternating façade was a new development, featuring an air intake system and a solar chimney for routing of the exhaust air.

4

Winter gardens at the National Museum of Science and Industry,
Parc de la Villette, Paris, RFR, 1986
With this hanging glazing, point-supported glass fixtures transfer the weight of the glass panes into the respective pane lying above; wire bracing absorbs the lateral forces.

Façade planning and construction

Façades are not limited to the actual space they occupy as part of the entire structure, but also influence the space in and around the building. A façade is the key element when observing a building from the exterior and has impact on the interior. View, lighting, ventilation, user comfort, some building services and possibly loadbearing are all tasks the façade may need to address. Façades are an integral element of the entire building with direct relation to design, use, structure and building services. This has decisive impact on the entire design and construction process.

Designing a façade is a process of communication and decision-making that focuses on the formulation of the building and its façade. The following steps can be defined as specific phases: initial conception, definition of functionalities, design, implementation coordination and assembly. Certain processes should occur during all of these phases: feedback on overall design and definition of functionalities as well as the element's importance within the overall structure of the building (structure, building services engineering, usage, safety).

We will describe an example of a functionality relating overall building design and façade construction. Water needs to be kept out of the building. The design might include, for instance, an overhanging roof with recessed windows. In terms of the construction process, a layered construction method with targeted drainage via eaves gutters or drainage edges would be preferred (5). Assembly would need to be executed from bottom to top in order to construct overhangs and ensure proper sealing.

5

Eave as weather guard, Ouro Prêto, Brazil
Use of a large eave as weather guard for the wall and window planes below. In the protected upper storey, casement windows were used, whereas in the lower storey, sliding windows were preferred since sealing this type of window is easier.

Similar relative effects can be seen in the example of a full glass façade: the design idea of a transparent envelope usually would entail the choice of a non-loadbearing façade to expose as much glass surface as possible (6). For the construction process, this would mean using a full glass façade set independently of the building structure. It would need to have a movable joint to avoid stress imposed on the glass façade by the main structure.

Why do we now emphasise on façades being highly technological components when they have always been part of the architect's scope of design? In particular, early Modern Architecture captivates us with its technically simple detail solutions. Single glazing could be made simply, without complex aluminium profiles, just flat steel welded for an extremely slim section.

Today's buildings stand in stark contrast to these historic examples of Mies van der Rohe (7) und Niemeyer (9) – they consist of numerous complex and interlinked technical solutions for the loadbearing structure, technical equipment and the façade. Individual design specialisations have evolved for each of these building components. The modern façade is a complex structure with numerous functions and complex technical realisation. When we look at the architectural façade solutions of early Modern Architecture, it becomes apparent that they were relevant in their time, but no longer fulfil today's requirements. Increased demands concerning comfort, e.g. heat-insulation as well as air- and rain-tightness, in the context of most industrial countries no longer allow the use of single glazing. This results in the need for thermal separation of the construction profiles followed by the consequent need to maintain this separation throughout window casements, drainage plains and jointing technology. The complexity of the technical aspects alone increases exponentially. If we further consider today's increased knowledge in material science and its rapid development, the possibilities seem endless, but so do the problems.

6

**Farnsworth House, Plano, Illinois,
Ludwig Mies van der Rohe, 1950**
This summer house lies embedded in the land-
scape; it sits on stilts to resist the annual floods
and to evoke a sense of detachment from the
surrounding area.

Development trend complexity

The prevailing trend in façade technology is its increasing complexity. The range of possibilities is expanding constantly and technical solutions become indicators for the state-of-the-art: more and more 'intelligent' façades (10) are being developed with the aim to increase the user's comfort level. But since users need to undergo a process of familiarisation with the new technologies, the question of their actual practicality remains partially unanswered. Thus, some developments are being revised and a few, however sensible with regards to substance, even disappear. Today we know of the issues of double façades and can better judge their advantages and disadvantages. They were built in quantities, but flawed conceptual designs or incorrect use and operation damaged their reputation. When looking at proprietary technologies and systems, we can see that these double façades do have historical predecessors; the Mediterranean box-window (8) for example, or decentralised air-conditioning units, so called 'fan coil units', which we have seen in older American high-rise façades.

7

Façade detail Farnsworth House, Plano, Illinois
The detail consists of an inner flat steel angle-bracket, a clamped single glazing without thermal insulation, and an outer finishing strip. As was customary in those days, no thermal bridges or drainage within the profiles were provided. Because the house was used during the summer months only, this was deemed unnecessary.

8

Historic façades in Bilbao
In Mediterranean climates, glazed balconies serve as part of the living space during transitional periods and as part climatic buffer during summer.

Current topics in façade technology development include energy, user comfort, individual façade expression (11) as well as adaption of existing façades. These topics are all driven by the search for new solutions to create façades for varying functions, climatic circumstances and geographic locations (12). The authors expect two major trends to develop: further emphasis on technical developments with improved design tools, manufacturing methods and system variants, as well as simplifying the façade by integrating components and functions into façades that might be complex to design but easy to manage.

However, exclusiveness does not exist in façade technology: there are no definite right or wrong solutions. Façades always result from individual creative conceptions, designed for a specific place, context and architectural concept. This book should be viewed as a guide to analyse, consider and develop. It challenges the reader to stay informed about new as well as conventional topics, to learn by observing, inquiring and visiting construction sites.

Façade planning is an integral part of the design process that employs constant feedback. It is a process based on progressive steps. This book is structured according to this schematic: the chapter 'From Wall to Façade' discusses the development of today's façades and their typological classification; the chapter 'Principles of Construction' explains the interrelationship between the building structure and the façade system; the chapter 'Detailing and Tolerances' broaches the issue of generating technical details for the general solutions defined previously; topics such as integrated design and building structure aspects of the façade are discussed in the chapter 'Climate and Energy'; the chapter 'Adaptive Façades' analyses how façades can adapt to changing parameters; the section 'Case Studies' describes typical and special façade solutions on the basis of selected projects.

9

Banco Mineiro de Produção, Belo Horizonte, Oscar Niemeyer, 1953
The slim sections and single glazing have remained unchanged in Oscar Niemeyer's administrative building because the local climate does not necessitate thermal insulation. Air-conditioning units for cooling the interior space in summer are positioned according to the requirement of the individual user.

10

ARAG Tower, Düsseldorf, RKW Architektur + Städtebau with Foster and Partners, 2000
This well designed double façade is a shaft-box system with individual box windows and cross-storey exhaust shafts within the glass façade.

Four new, exemplary case studies were selected for this second and revised edition. They explain four different construction methods that are in common use; particularities are noted in the text. The case studies are meant to be transfered to a particular project or problem. Thus, the rear-ventilated façade of the Concept House in Rotterdam, the Netherlands, is an interesting example of façade cladding that could be similarly realised with many different materials in sheet or panel form. The second case study, the State Archive Nordrhein-Westfalen in Duisburg, Germany, deals with a more traditional single-skin solid masonry construction, which is particularly helpful as an example for extensions to older buildings or restoration of historic buildings. Both examples are wall constructions with windows, and therefore so-called punctuated façades.

The desire for transparency and façades with large glazed areas is addressed with the other two case studies. The new building for the Department for Architecture and Interior Design at the University of Applied Sciences in Detmold, Germany, was executed with a post-and-beam façade; a suitable solution for buildings only a few storeys high. The element façade of the Headquarters Süddeutscher Verlag in Munich is a good example of a very modern high-rise façade, delivered as prefabricated elements and assembled storey by storey. Following the punctuated façades of the examples in Rotterdam and Duisburg with their loadbearing walls, these two curtain wall variants offer an informative introduction into the design of façades which are placed in front of the loadbearing structure of the building. In closing, the authors offer an outlook on the future of façade technology and the possible lines of development.

11

**Arab World Institute, Paris,
Jean Nouvel, 1989**
South façade of the Arab World Institute with a technical interpretation of the Arabic sun screen as an integrated-pane system. The blinds open and close depending on the angle of the sun.

12

**Juscelino Kubitschek Complex,
Belo Horizonte, Oscar Niemeyer, 1951**
North façade of a residential high-rise building from the fifties with sun protection lamellas which can be adjusted for each flat individually. Lamellas of varying incline create a textured surface that changes the building's appearance – from a design point of view, a very modern façade.

2 | From Wall to Façade

The form and function of present-day wall and façade constructions are the result of a long process of development, which is closely related to the history of humanity. Starting from the two original basic forms of human existence – the settled and the nomadic – and the functional, technical and design-related requirements resulting from these conditions, we can outline the resulting forms walls and façades take and their further development. Depending on climatic conditions and the various life styles and dwelling styles that grew out of them, two essentially different basic principles for the construction of the outer envelope of a dwelling place came into use: solid walls fixed to one particular spot and designed for permanence on the one hand, and more flexible, less permanent façades – typically represented by tents for mobile use – on the other.

The survey of the development trajectory given here follows not so much cultural or historical as construction trends against a background of structural and functional relationships. Thus, the development is not chronological but one in which the successive steps of the construction developments involved to bring out the interdependencies and relationships inherent in them, as well as the underlying logic. The resulting overview of the phenomenon of the façade should be seen as a snapshot from current perspectives that understandably focuses on present-day developments but not limiting itself entirely to them.

Solid wall construction

People who lived in cold climates and populations who had adopted a settled mode of life preferred wall constructions that were as solid as possible (1, 4). Such walls are built up either of readily available building materials or of elements made suitable for the purpose by simple processes, such as naturally occurring stones, squared stone or fired bricks. The objective was to build a wall that would stand up to climatic influences while still keeping the building method as uncomplicated as possible. Though the construction and finishing of such solid structures has naturally developed in line with advances in technology – present-day solid walls are either built up of structural units with both loadbearing and thermal insulation properties or are provided with elements for this purpose – the basic principle remains unchanged.

Warm façade, cold façade

Two different types of solid wall construction may currently be distinguished: warm façades (2), where the insulating layer is mounted directly on the outside or the inside of the façade construction, and cold façades (3, 5) where the insulating layer is separated from the climatic protection layer by a layer of air. The latter principle allows the insulating layer to dry out if water penetrates into the façade as a result of damage to the protective layer.

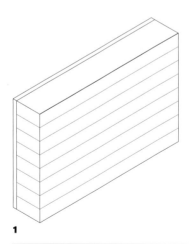

1

Solid wall
Solid wall constructed from monolithic or composite elements. Often, the masonry is plastered, shown here with an internal plaster.

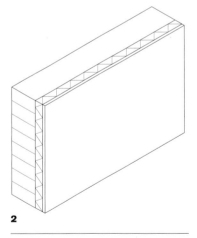

2

Warm façade
Warm façades have a thermal insulation layer applied directly to the surface of the building. If the insulating layer is applied on the outside, it also has to be water-resistant to ensure that the insulating properties are not lost due to weathering. If the insulating layer is on the inside, the ability of the solid wall to store heat will no longer actively influence the interior environment.

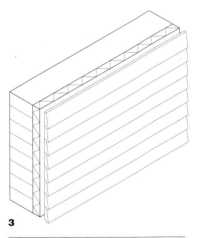

3

Cold façade
Cold façades are characterised by the presence of a cavity, ventilated internally, between the outer layer that offers protection against the weather and the thermal insulation layer. Potential condensate can escape via the cavity.

4

Marketplace in Siena, 13th century
Solid masonry wall as a loadbearing and space-enclosing structure.

5

**Port Event Center, Düsseldorf,
Norbert Wansleben, 2002**
The cold façade of the ground floor of an office building in the port of Düsseldorf. The transparent outer layer that offers protection against climatic influences allows the skeleton of the building, the ventilated cavity and the thermal insulation layer to be seen.

Openings in solid wall construction

Openings were originally made in the walls to allow smoke to escape (6). At a later stage in the development, the openings were enlarged to let light in. The method used initially to solve the problem of the weakening of the fabric of the solid wall by the creation of openings in it was to use horizontal beams as lintels.

In Gothic architecture, the amount of solid masonry used in the wall was gradually reduced to allow large areas of glass to be incorporated into the walls, with the aid of constructive techniques of which the ingenuity is still impressive today.

Driven by the wish to admit even more light into the interior of the building, the amount of masonry used in the wall was gradually reduced (8). As the Romanesque style of architecture was succeeded by the Gothic, the previously almost monolithic walls were progressively replaced by filigree structures that may be regarded as precursors of the skeletons used in present-day building techniques. The roofs were shell constructions with cross-wise support, resting on pillars and loadbearing walls (7). This allowed the vertical forces to be concentrated at a number of predetermined positions, from where these forces were transferred to the ground. This made it possible to create large openings in the relatively unstressed parts of the walls. Since the transfer of loads also leads to lateral forces in this system, these lateral forces also have to be transferred to the ground by means of suitable ties or external buttresses.

6

Dissolution of masonry in a church window
The introduction of large windows in churches and cathedrals went hand in hand with a reduction in the area of masonry. The glass area was divided into small window frames.

7

Cathedral of Amiens, 1220-1269
The space was opened up by dividing the
structure into loadbearing and covering elements.
Large areas no longer had any loadbearing
function, thus allowing windows to be created
in them.

8

Merchants' houses in Antwerp, 16th century
As the merchant classes increased in stature,
building methods permitting large areas of glazing in façades also came to be used for profane
purposes, as in these historic merchants' houses
surrounding the marketplace in Antwerp.

Bridging the gap

Since the use of window lintels to span large openings in walls (9) very soon reached its structural limits, the next stage in the development was the use of arches for this purpose. In Gothic architecture, these arches became pointed since this form is more capable of bearing the weight of the wall lying above. Present-day building styles continue to use lintels – made of steel or reinforced concrete – to span openings in solid walls (10).

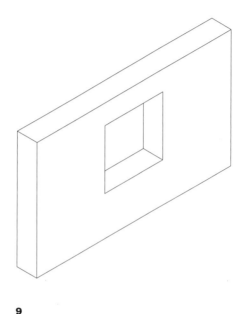

9

Openings in walls
Openings in a solid wall allow fresh air and light access to a building.

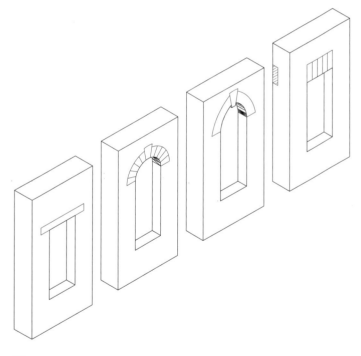

10

Bridging the gap
Openings in solid walls were initially spanned with wooden lintels. Arches were later introduced, since they made it possible to bridge wider gaps. In the Gothic style, these arches were pointed to allow them to bear greater weights of masonry above them. Modern architecture would conventionally make use of concealed loadbearing elements made of steel or concrete to span large openings.

Closing apertures – single glazing

In addition to creating apertures in the wall to introduce light and air into the room there is the requirement to close off the space against cooling and unauthorised access. Thus, the apertures were filled in with translucent materials – in the case of Roman thermal baths, for example, with thinly cut sheets of marble. The development of glass as a building material made it possible to fill the openings in the walls with single panes of glass that not only provided natural lighting in the houses, but also allowed the people inside to view out (11, 12). The production technology initially only allowed small panes of glass to be made, and the resulting windows were correspondingly small. Development of the glass-in-lead technique made it possible to construct much larger windows. This, combined with the use of stained glass, allowed magnificent results to be achieved especially in sacred architecture. At present, large single glazing is often used mounted in steel frames (10).

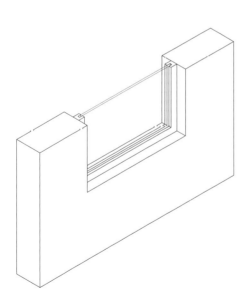

11

Single glazing
Single glazing was used in apertures to provide natural lighting, to allow views out from inside the building and vice versa, and to prevent heat loss from the building. Various techniques were developed to join initially available small panes of glass so that larger openings could be filled.

12

Weißenhof Siedlung, Stuttgart, 1927, Ludwig Mies van der Rohe, Le Corbusier, Walter Gropius
Single glazing in steel window frames in a building complex at the Weißenhof Siedlung in Stuttgart. The windows are positioned on the outside of the wall apertures to produce a flat façade.

Box window

The box window may be seen as a further stage in this development. Here a second pane of glass is added, slightly set back from the first, to create an additional climatic buffer if this is required by the climate or the season (13, 14). The space between the panes is not hermetically sealed, to avoid condensation (15).

This may be called the first intelligent wall: depending on the state of the weather or the occupant's needs, the second window pane can be slid up or opened otherwise to let in the outside air or slid down to improve thermal insulation. Depending on the climatic conditions, the user decides how many panes should be opened or closed.

14

Box window
Depending on the season, the outer window sashes can be opened or closed. Shown here are summer conditions; the configuration allowing free window ventilation without overheating of the cavity between the inner and outer window panes.

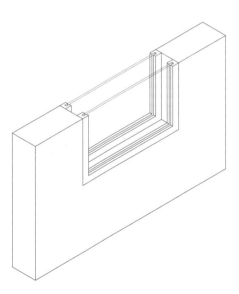

13

Box window
A box window is produced when a second pane of glass is installed to meet seasonal conditions. Depending on the temperature, the user can decide how many panes should be opened.

15

Loggias in façade, Bilbao
These loggias built in front of the actual windows may be regarded as a variant of the box window: in spring and autumn, these spaces can be used as an additional room while in winter they are closed off to act as an extra climatic buffer.

Insulated glazing

The next step arguably in the development of the box window is insulated glazing or double glazing. This consists of two panes of glass permanently joined together with an insulating layer of air or inert gas in between (16, 17), providing a more effective barrier between the internal and external environments.

After some experimentation with different methods of glass mountings, the panes of glass are nowadays usually mounted in aluminium or plastic profiles with the aid of a silicone sealant.

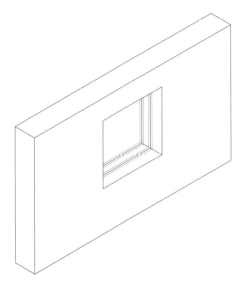

16

Insulated glazing
Two panes of glass permanently joined together to give insulated or double glazing a more effective barrier between the internal and external environments.

17

Fondation Cartier, Paris, Jean Nouvel, 1994
The façade consists of single glazing on the left and double glazing on the right.

Walls with skeletal structure

In a parallel line of development found in nomadic societies, tents consisting of a supporting skeleton and an outer covering to keep out the elements were the main form of building (18). It would be clear that in order to facilitate the transport of the tent, both the skeleton and the outer cover had to be as lightweight as possible. There was no place for massive structural elements here. Under these conditions, it was necessary to separate the functions of support and enclosure.

Half-timbered construction

The two parallel developments outlined above, of the gradual dissolution of the solid wall to give more window space, and of the tent with its separation of support and enclosure, combined to bring about the gradual transformation of the solid wall into the relatively lightweight modern façade. This is achieved by building a supporting frame or skeleton (originally of timber) and infilling in the intermediate spaces with an appropriate cladding (20). The European predecessor of this building technique is half-timbered construction (19), in which a timber skeleton is built and the spaces in between the timber elements are infilled in with different materials according to the region: interlaced branches, mud or clay (the combination of these two being the well-known 'wattle-and-daub'), or bricks. Apart from the choice of material for infilling the gaps in the timber skeleton, these structures may vary in the method used to mount successive stories: the ceiling of one storey, and the floor of the one above, may either be mounted in the wall (as in the French or Normanic tradition) or rest on the wall (as in the Germanic tradition). In the latter case, we get the overhanging of successive storeys that was characteristic of medieval European buildings.

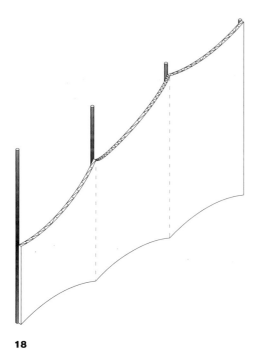

18

Tent
Tent-like structures arose through the need to move one's home frequently from one place to another as is inherent in the nomadic way of life. These structures were designed for ease of assembly and dismantling, which made it necessary to separate the functions of support and enclosure from each other.

19

Half-timber houses, Detmold, 16th century
In a half-timbered construction, the loadbearing capacity is provided by the skeletal structure while the infill in the intermediate spaces merely has an enclosing function. The overhanging of successive storeys is clearly visible.

Platform and balloon framing

The American variant of this principle is the timber-frame construction. This consists of bearing timber members, the spaces between which are infilled with sheets of wood products. Since this type of construction does away with a solid outer wall, it has poor thermal buffering properties.

A distinction is made here between platform frames and balloon frames: in balloon frames (22), the ceiling of one storey, and the floor of the one above, are mounted in the wall while in platform frames (21) they rest on the wall. In multi-storey buildings, the walls are erected on the completed platform provided by the flooring.

21

Platform frame

In the platform frame variant of the timber-frame construction, widely used in America, the walls are made of vertical timber members of which the spaces in between are infilled with wooden boards and thermal insulation material. The ceiling of one storey, i.e. the floor of the one above, rests on the wall.

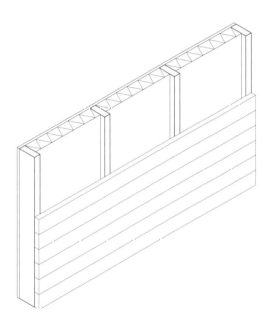

20

Timber-frame construction

A timber-frame structure is built up of relatively slender timber members, the space between which is infilled up with cladding on the inside and the outside. The space between these two layers of cladding acts as thermal insulation.

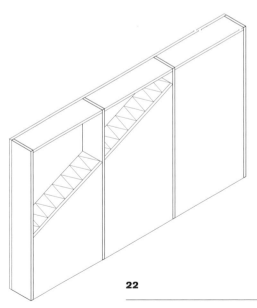

22

Balloon frame

A balloon frame consists of posts one storey high provided with cladding on the inside and the outside. In contrast to platform frames, balloon frames have the ceiling/floor unit built into the wall.

Resolution of the wall into loadbearing structure and façade

The development described above took thousands of years. The subsequent steps in the development process occupied no more than about a century. Once again, there is no need to trace the development in chronological order – it makes more sense to outline the structural development trajectory.

In the run-up to the Neoclassical era, architects did their best to separate the various functions of the wall even further. Bearing, sealing and the transmission of light were becoming more and more clearly distinguished from one another, though it may be noted that technical limitations did not yet allow the loadbearing function to be completely separated from the others. Nevertheless, it had become possible, for example, to incorporate large window openings in the wall without the need for structural connections of the kind that were required in the churches and cathedrals of previous centuries.

On the basis of these developments, Neoclassical architects finally succeeded in separating the outer envelope of a building completely from its loadbearing structure, thus allowing the wall to dissolve into a façade. The bearing function is provided by columns, which are as far as possible enveloped into the interior of the building, while the façade leads an almost independent existence on the exterior (23).

23

**Farnsworth House, Plano, Illinois,
Ludwig Mies van der Rohe, 1950**
An example of the total separation of loadbearing structure and façade is provided by the Farnsworth House designed by Mies van der Rohe. The plane of the loadbearing structure is situated in front of the façade, emphasising the dual-plane design theme. The glass façade situated behind the loadbearing members is very unobtrusive, almost eliminating the visual difference between inside and outside in this highly innovative creation.

Post-and-beam façade

The further resolution of the wall into façade and loadbearing structure is based on theoretical ideas that were developed at the time, and led in the last analysis to the glass boxes that are such a common feature of present-day cityscapes.

The next evolutionary stage – not so much in chronological as in structural order – was the development of the post-and-beam façade as the logical next step in the dissolution of the solid outer wall. This system consists of storey-high posts linked by horizontal beams. The gaps between successive posts and beams can be made to perform various functions, such as for cladding, lighting and ventilation (24, 25). In these standing post-and-beam façades, the posts serve not only to transfer the wind forces and self-weight of the structure to the ground but also to provide support for the cladding and other functions.

24

Post-and-beam façade
Post-and-beam façades consist of storey-high vertical posts linked by horizontal beams. The spaces between these members house the appropriate functions.

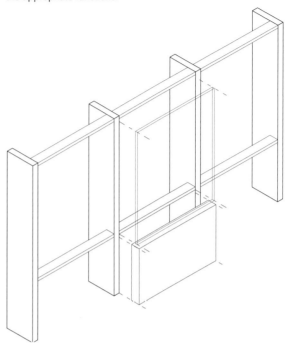

25

Library, Delft University of Technology, Mecanoo, 1998
Post-and-beam system used in the new library at Delft University of Technology, consisting of vertical post and horizontal beam elements. The panes of glass are mounted in external frames.

Post façade

Apart from pure post-and-beam systems, post systems and beam systems (where tie rods are used to bear loads) have also been developed. The objective of these variants is always to increase the degree of openness, thus improving the transparency of the structure. In post systems, the structural limit is determined by the maximum permissible distance between posts (26).

Beam façade

When the construction is reduced to using only beams instead of a standing post-and-beam system (29), the result is a suspended system (27) where loadbearing capacity comes from above to reduce structural mass and to avoid exposing structural elements to buckling. Such systems generally require heavy-duty tie-rod structures mounted near the roof to bear the entire weight of the façade. The beams only have to resist lateral forces.

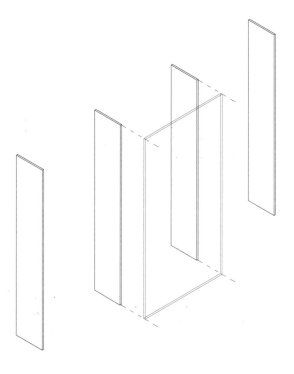

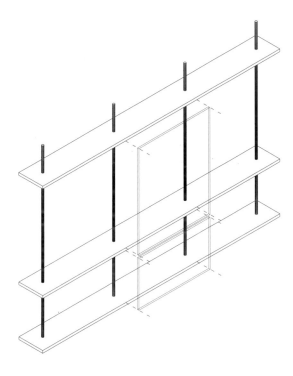

26

Post construction
The storey-high posts lead wind loads and the self-weight of the structure to the ground.

27

Beam façade
Façade constructions in which only beams are used require a vertical suspension system to bear the weight of the façade. Wind loads are here transferred to the ground via the beams.

Curtain wall

From a structural point of view, systems in which the façade hangs from the front of the roof may be regarded as precursors of the development of curtain walls (28, 30). Since the construction is practically independent of the building's main loadbearing structure, the façade can be partitioned almost at will and cladding or glazing used to meet the various aesthetic or functional requirements. The vertical and lateral loads are generally led to ground floor by floor, but special loadbearing elements may be added to bridge longer spans.

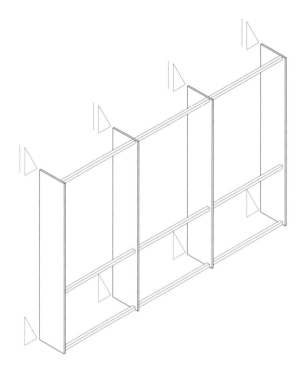

29

Standing post-and-beam façades
Standing post-and-beam façades usually consist of storey-high modules. However, the problem of the post being subjected to buckling has to be considered.

28

Federal Center, Chicago, Ludwig Mies van der Rohe, 1964
Mies van der Rohe's Federal Center in Chicago is an example of a curtain wall. It reflects the demand for industrially produced façades that at the same time satisfy architectural preferences: the façade is made up of prefabricated elements, assembled by craftsmen on site.

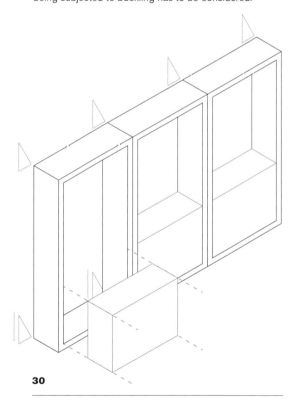

30

Curtain wall
Unlike pure post-and-beam systems, curtain walls are suspended from above with the aid of tie rods. This approach has the advantages of avoiding buckling in the posts and of a large degree of independence from the main structure of the building.

System façade

The curtain walls used at the moment may be divided into stick and unit systems (31). Depending on the type of construction, it may be possible to prefabricate various elements of the wall and assemble them on site, or to prefabricate the entire system wall off-site and install it as a whole. Its advantages are a guaranteed production quality, rapid assemby and low labour requirements on site. Unitised systems have become particularly established in the area of high-rise façades (32), since here units are used in a highly repetitive manner and can be mounted from the storey above with the aid of a crane. For lower buildings with easy access around all sides of the building the consideration should be whether a unit system façade offers advantages in terms of the building process and progress.

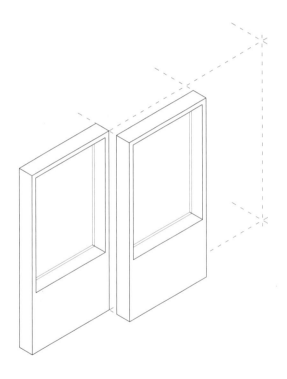

31

System façade
System façades, unlike post-and-beam systems, can be fully prefabricated and mounted on site by a small labour force.

32

Westhafen Haus, Frankfurt, Schneider + Schumacher, 2005
This system façade consists of storey-high façade elements mounted on an external frame, together with glazing and ventilation ducts built into the vertical posts. The separation of the façade from the building's loadbearing structure by means of the supporting elements behind the façade is distinct.

Double façades

One of the interesting developments that may be observed at this time is the rise of the double façade (33) resulting from the shift of various functions related to the interior functions of the building immediately behind the façade. For example, instead of installing ventilation systems in the building, the ventilation can be provided by thermal insulation between the two layers of the façade. On the basis of experience, the initial variants of this concept have developed into ventilation systems encompassing one or several stories. The initial euphoria concerning the techni-cal and design possibilities of this approach has been replaced by a more pragmatic approach on the basis of greater under-standing of the mechanisms underlying the effects produced. As a result, double façades now tend to be used only when called for. There is no need for double façades on all sides of every new building. Under certain circumstances, however – e.g. high levels of street noise, high wind loads or increase in building height – such façades may be the appropriate and economical solution (34).

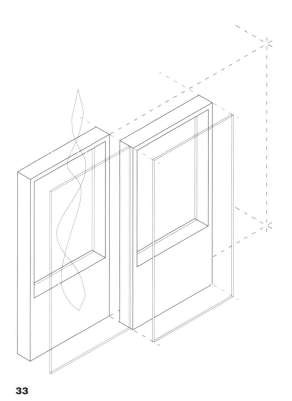

33

Double façade
A double façade is obtained by adding an extra layer of glazing outside the façade to provide the building with ventilation or additional sound-proofing. This system may be realised in various ways, depending on the functions desired and the requirements made on the façade.

34

Single and double façade:
Triangle Building, Cologne,
Gatermann + Schossig, 2006
An example of the façade for a high-rise building offering different functions depending on the requirements. The single façade may be seen on the right of the picture, while on the left an ad-ditional layer of glazing has been added to create a double façade with a ventilated space between the two layers.

Second-skin façade

On the basis of present knowledge of the underlying principles, four main types of double façades may be distinguished. The first variant, known as the second-skin façade, is obtained by adding a second layer of glass over the entire outer surface of the building (35, 36). This has the advantage of technical and structural simplicity and the fact that it does not involve a large number of moving parts since the outer single layer of glass is simply mounted on the inner insulated glass façade structure. The disadvantage is that it offers few possibilities of controlling the interior environment of the building; there is thus an attendant risk of overheating.

Box-window façade

The second variant embodies the above-mentioned principle of the box window, by including storey-high façade elements in the system, which individual users can open at the top and the bottom (37). The advantage of this model is the freedom the system gives individual occupants in controlling their own internal environment. The disadvantage is that the freedom given to one occupant may have an adverse effect on the conditions experienced by another, since e.g. the exhaust air from one floor can influence the quality of the incoming air on the floor below. This problem can be avoided by staggering the ventilation inlets and outlets.

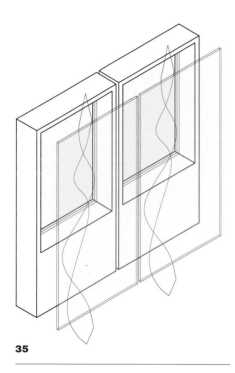

35

Second-skin façade
A second-skin façade is produced by adding an external layer of glass to the inner façade. This has the advantage of being easy to construct but the disadvantages of limited control possibilities on the interior and, in the case of high buildings, the attendant risk of overheating.

36

Box-window and second-skin façade
On the left we see a window element added on the inside to form a box-window façade, while on the right an early example of a second-skin façade may be seen. This has been created by adding an additional layer of glass outside the basic façade.

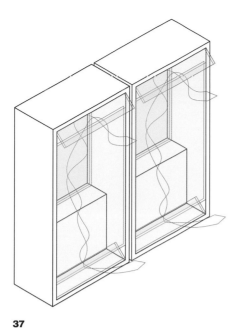

37

Box-window façade
Storey-high box windows with ventilation flaps at top and bottom offer the possibility of individual control.

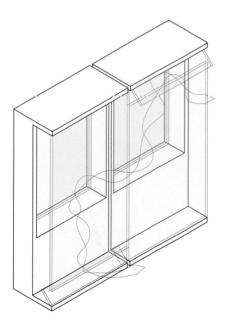

38

Corridor façade
Corridor façades connect neighbouring double-façade elements in order to permit staggered ventilation of the space between the two skins.

Corridor façade

To deal with the problem of interference between the ventilation systems at different levels, the third variant – the corridor façade, with staggered air inlets and outlets – was developed. This used horizontal baffles at ceiling height in the space between the two skins. However, the uninterrupted horizontal flow of air could give rise to noise interference between neighbouring rooms. It is not always possible to install these baffles later, since this type of façade depends on the presence of horizontal connections (38, 39). Particularly due to the uninterrupted space between the two skins, the building can be naturally ventilated from all directions.

39

Stadttor Building, Düsseldorf, Petzinka Pink und Partner, 1998
An early example of a corridor façade: the storey-high façade elements have rotary timber baffles with insulated glass on the inside and a continuous single-pane glass skin on the outside.

Shaft-box façade

The most effective version of the double façade, but that involving the greatest constructional and control-engineering effort, is undoubtedly variant number four, the shaft-box façade (40, 41). Discrete box windows or other façade elements release their exhaust air into a vertical shaft mounted on the façade and extending over several floors for greater thermal efficiency. The height of the shaft means that a stack effect ensures vertical motion of the air in the shaft, hence enhancing the efficiency of the system. However, in order to allow for controlled ventilation, an adjustable ventilation flap leading to the shaft is required in every box window element.

41

Photonics Centre, Berlin,
Sauerbruch Hutton Architects, 1998
Early variant of the shaft-box façade, consisting of vertically separated ventilation shafts in the plane of the façade which merge at the top for effective ventilation of the space enclosed by the double façade.

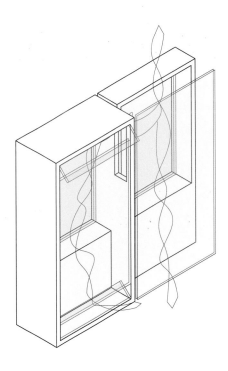

40

Shaft-box façade
Shaft-box façades, featuring box windows that release their exhaust air into a shaft that extends over several floors, offer a double façade system that requires complex installation but is highly effective.

Alternating façade

The double façades described above do not offer complete solutions to the problem of variable ventilation requirements. One approach to this problem was the development of alternating façades, also known as hybrid façades (42, 43). These are basically single-skin façade constructions that can be converted locally to double façades by the addition of a second skin. The objective here is to combine the benefits of the simplicity of the single-skin façade with the buffering effect of the double façade. In summer, ventilation can take place via the single skin area of the façade; an additional exterior grating can provide efficient ventilation during rain or at nighttime. In winter, the double-skin area of the façade can be used for ventilation by using the warmed up air from the cavity between the two skins to ventilate the interior. During summer, this area can be opened to prevent overheating of the cavity.

43

**Debitel Headquarters, Stuttgart,
RKW Architektur + Städtebau, 2002**
RKW worked together with Transsolar Climate Engineering to develop an alternating façade for the Debitel head offices in Stuttgart. Different parts of the façade in this building were built as single-skin façade with a permanent louvre layer, single-skin façade with a louvre layer behind it and double façade.

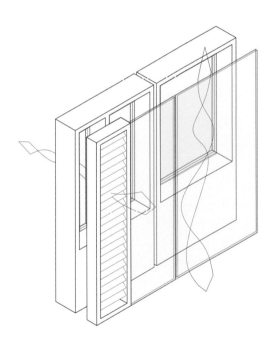

42

Alternating façade
In alternating façades, a second skin is added locally to a single-skin façade construction to provide the benefits of the buffering effect of the double façade in the areas affected. A grating can be mounted in front of the single-skin areas to allow for ventilation during rain and at nighttime regardless of weather conditions.

Integrated façade

The idea of the double façade underwent consistent further development by integrating functions other than ventilation, such as air-conditioning or control of lighting levels, in the façade. The resulting system was then generally called a 'modular façade' or 'hybrid façade' (44, 45). When taken to the extreme, it offers the possibility of divesting the building itself of all functions apart from that of bearing its self-weight and incorporating the enclosure function as well as all environmental-engineering functions within the façade. This constructional approach could thus engender a hitherto unknown synergy between façade construction and internal environmental control engineering, leading to a fundamental change in building design. Instead of the old core-oriented approach, a number of essential functions are now transferred from the core to the façade. Several such concepts have been developed by system suppliers to date; however, due to the necessity to make according decisions very early in the design process, the market is hesitant to employ these systems, even though the high degree of prefabrication of integrated façades makes them advantageous.

44

Post Tower, Bonn, Helmut Jahn, 2003
Helmut Jahn worked together with Transsolar Climate Engineering to develop one of the first hybrid façades for the Post Office Tower project in Bonn. Environmental-control modules built into the top part of the façade could be controlled locally as individual units.

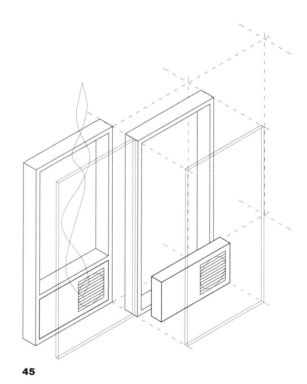

45

Integrated façade
The integrated façade incorporates not only ventilation functions as described above but also active environmental-control or lighting components.

During the design of the Lloyd's building in London in 1978 Mike Davies, a colleague of Richard Rogers, developed the concept of the 'polyvalent wall' – a façade that apart from the classic functions of sealing and insulation can also assume other functions such as environmental control, ventilation and individual control of lighting. Consideration of current developments in the hybrid façade shows that we still have a long way to go to reach the ideal polyvalent wall envisaged by Mike Davies. Apart from the above-mentioned limitations associated e.g. with the double façade (46) and the technical problems associated with the construction, operation and logistics of the modular façade, it may be argued that concentrating exclusively on improving the glazing is a dead-end approach: if one tries to incorporate all functions in the glass, problems still remain in the design of a particular building component or the choice of the best building material. At present, a more sensible philosophy would seem to be the separation of functions into various levels and their incorporation within various building components, which are then ultimately combined in the modular façade (47).

46

**Stadttor Building,
Petzinka Pink + Partner, 1998**
An early example of a double façade: the exterior glass envelope protects the interior timber façade.

47

debis Headquarters, Renzo Piano, 1997 and Daimler-Chrysler Building, Hans Kollhoff, 1999, Potsdamer Platz, Berlin
The appearance of these façades – glass or solid – is quite different while function and use are similar.

3 | Principles of Construction

The façade separates the interior from the exterior. Before addressing today's façade constructions we would like to call to mind the different functions that a façade serves: it defines the architectural appearance of the building, provides views to the inside and outside, absorbs push and pull forces from wind loads, bears its self-weight as well as that of other building components. The façade allows sunlight to penetrate into the building while usually providing protection from the sun at the same time. It resists the penetration of rainwater and has to handle humidity from within and without. The façade provides insulation against heat, cold and noise and can facilitate energy generation. In addition, it must be long-lasting, easy to maintain and to clean.

Sketch 1 shows the complexity of the requirements to be fulfilled. These requirements need to be considered during all phases of the façade construction: during the conceptual phase, while working on the principles of construction, during detailing and lastly during construction and usage.

Basically we desire a structure that is as simple as possible yet carries out all these functions and is adaptable to changing influencing factors. It should be an adaptive envelope similar to the human skin, fulfilling several functions of the body.

Today's façade is based on developments spanning several millennia. The solutions currently in use result from tried and tested construction methods, the materials available and traditional production and assembly processes.

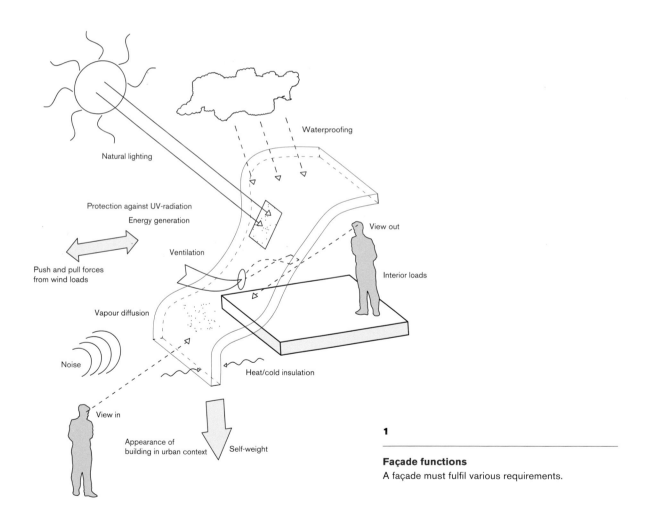

1

Façade functions
A façade must fulfil various requirements.

Areas of construction

In the following we will describe the principles of construction using a metal and glass façade as an example.

Three main areas of construction (2) can be defined within the façade:

• Primary structure (shell of building) forming the main loadbearing structure of the building
• Secondary structure, which is the loadbearing structure for the façade and constitutes the connecting element between levels one and three
• Infill elements

The primary purpose of this assembly lies in the separation of the above mentioned functional requirements that the façade needs to fulfil. The functions are distributed among several different components. This arrangement simplifies the connection of individual façade components with each other and provides options to compensate for movement due to wind as well as thermal expansion.

The primary structure takes on the loadbearing function of the entire building and transfers the loads from the façade to the foundation.

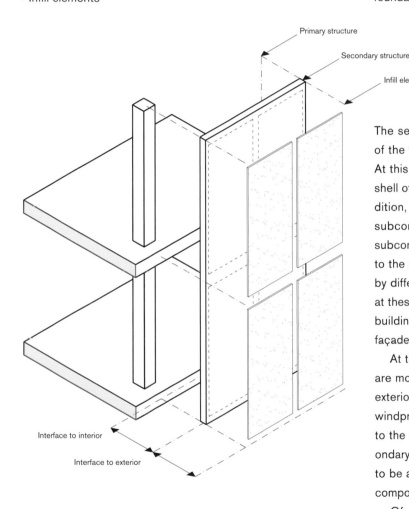

The secondary structure comprises the loadbearing structure of the façade. It transfers its loads onto the primary structure. At this 'interface to the interior' the differing movements of the shell of the building and the façade need to be balanced. In addition, these two structures are typically assigned to different subcontracts; the shell of the building usually falls under the subcontract for concrete work whereas the façade is assigned to the metal subcontract. As these elements are manufactured by different companies, there is a need for special coordination at these interfaces. Manufacturers' tolerances of the shell of the building (concrete) lie within the centimetre range whereas the façade (metal) tolerates only deviations of millimetres.

At the same time infill elements such as glazing, panels etc. are mounted on the secondary structure. This 'interface to the exterior' has to fulfil its own functions: the elements have to be windproof, resist water penetration, or it must be re-channeled to the exterior, movements between the elements and the secondary structure have to be tolerated and thermal bridges have to be avoided. Thus the secondary structure is a very complex component.

Of course there are also façade constructions where the primary and the secondary structures form one component, i.e. the secondary structure is part of the loadbearing structure of the building. In this case the interfaces to the internal and the external are reduced to one. When using this type of structure we need to closely examine it with regard to tolerances, deflection and building physics. And if the façade is part of the loadbearing structure of the building, individual façade components cannot be easily exchanged.

2

Schematic representation of the elements of façade construction
In principle all façade constructions are based on this schematic design; however, different functional requirements can be combined into one component.

The actual space enclosure is created by the infill elements (3). These can comprise glass panes for lighting and view, panels for heat insulation and opening flaps for ventilation. The elements can also be layered. For example, it might be practical to arrange sun shading on the outside of the glazing, and glare protection on the inside. Double façades are another example of the principle of layered functions. Basically all façade designs can be categorised according to this system.

Façade bearing structures and load transfer

We can differentiate between different types of loads affecting the façade structure:
• Self-weight of the façade components
• Weight of snow
• Wind load (push and pull)
• Live loads e.g. a person colliding with the inside of the façade which in turn must be prevented from falling. (Fall protection)
• Stress loads. These are caused by deflections of components through changes in temperature or humidity.

3

Academy Mont Cenis, Herne, Jourda & Perraudin, 1999
The timber column of the primary structure can be seen on the inside; the secondary structure consists of wooden posts. Glass panes and ventilation flaps form the space enclosure.

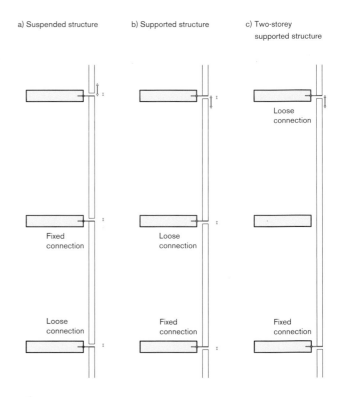

a) Suspended structure   b) Supported structure   c) Two-storey supported structure

Loose connection

Fixed connection   Loose connection

Loose connection   Fixed connection   Fixed connection

4

Load transfer
The drawing shows a suspended structure on the left, a single-storey supported structure in the centre and a two-storey supported structure on the right.

First we will examine a façade with extensive glazing and a perimeter frame structure (5) and its load transfer as per the above mentioned areas of construction.

The self-weight of the glass panes acts downward parallel to the façade. If glass panes are not fixed by planar fittings or suspended from above they sit on two plastic blocks. Only two blocks are necessary because vertical glass panes do not sag and therefore rest on two support points only, no matter how many support blocks are being used. For optimum structural integrity these blocks should be situated at a fifth of the distance from the edge. In this example the load bears on the lower edge of the frame (secondary structure). Depending on the type of glass used, a functional panel with several individual glass panes can weigh half a ton and more! The fixed connections hold the façade in place whereas the loose connections compensate for movements of the construction.

The push-pull-forces from the wind load and other dynamic loads acting vertically on the façade are transferred from the functional layer to the linear secondary structure (frame).
In turn the secondary structure transfers the loads to the primary loadbearing structure. It is to be expected that the façade as an exterior building component is subjected to different weather conditions than the shell of the building on the interior. Additionally, the façade is usually made from different materials, resulting in different linear expansions. Furthermore the primary structure is impacted by other loads and is liable to deflect accordingly. To avoid wedging, the secondary structure is supported on its lower edge or it is suspended from above (4).

In most cases it makes sense to transfer the façade loads storey by storey and to add expansion joints so that variations in dimensions do not add up across several storeys. This becomes possible when the secondary and the primary structures are separated.

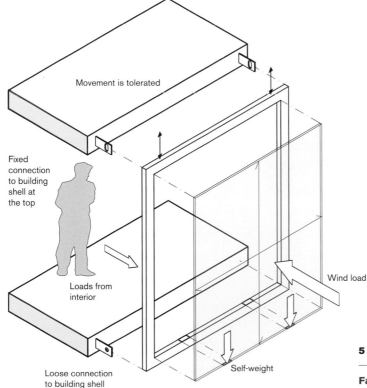

Movement is tolerated

Fixed connection to building shell at the top

Loads from interior

Loose connection to building shell at the bottom

Wind load

Self-weight

5

Façade load transfer
Different types of loads that need to be transferred.

If the façade is part of the primary structure (shell of building) and has a loadbearing function we need to consider the expansion differentials. In comparison to two separate structures, this can be a lot more complicated because the loads cannot be transferred via expansion joints as described above.

The drawing shows further examples of loadbearing systems for metal and glass façades (6).

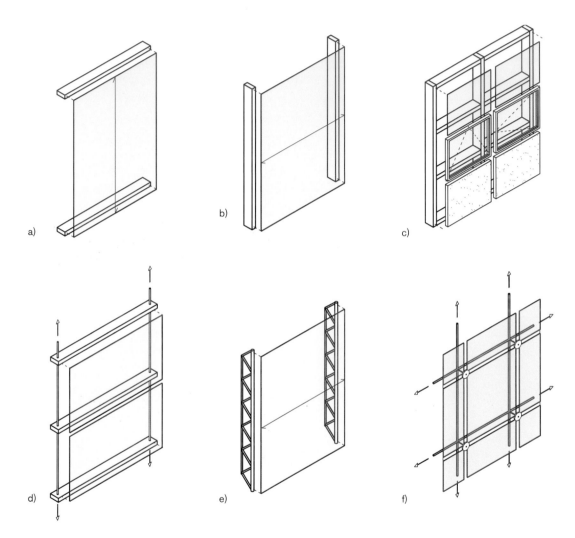

6

Loadbearing systems

a) Secondary structure without posts. The glass pane is subjected to one-sided vertical tension and must be dimensioned accordingly.

b) Secondary structure without beams.

c) Secondary structure with small partitions. Elements of varying functions are infilled.

d) Secondary structure consisting of lateral tie rods to transfer wind loads. Cables transfer loads up through the primary structure.

e) Replacement of the secondary structure by half-timbered structure.

f) Cable-mesh structure. The glass panes are connected with the cable-mesh structure by planar fittings mounted at the corners. The façade behaves like fabric under tension and relatively large linear deflections occur. This results in a rather large movement of the edge of the façade towards the façade plane, which has to be taken into account in the design of the structural connections.

In conclusion, there are numerous structural systems available for façade construction (7-9). The decision for a specific system depends on the following factors:

- Type of primary structure or shell of building
- Load transfer from the exterior towards the interior
- The size and properties of the infill elements (glass dimensions, deflections, weight, etc.)
- Architectural design

8

Wilhelm Lehmbruck Museum Duisburg, Manfred Lehmbruck and Klaus Hänsch, 1964
The façade consists of suspended glass panes with upper and lower framing. The suspended design allows for thinner glass material than a supported structure.

7

Detail of a cable-mesh façade
The glass panes are connected with the cable-mesh structure by point fixings mounted at the corners.

9

Point fixing façade
The loads from the glass panes are transferred via point fixings.

Grid and positioning of the façade within the building

Most buildings are designed with repetitive units, the so-called modular unit. The resulting grid helps to structure and organise the building volume into units based on the modular dimension. Thus, the position of each building component is specified and geometrically related to the adjacent components. Such grids are used to organise the entire floor plan as well as the individual components, e.g. masonry.

This kind of repetition is beneficial for the entire building process. In structural engineering, for example, a breakdown into standardised spans saves time and effort. Planning and communicating with project team members is simplified when everything relates to a basic grid. Even the furnishing of a building becomes easier. Of course different types of buildings may require different grids due to particular requirements, resulting in different structural systems.

Office buildings, for example, are usually based on a grid of 1.35 m, allowing efficient furnishing. If the building comprises an underground parking garage the primary structure is usually based on a structural grid of 5.40 m or 8.10 m, both multiples of the 1.35 m modular unit. This leaves sufficient space between the columns for two or three car-parking spaces.

It is most efficient to use the same grid for the façade as for the building itself. Typically we distinguish between a primary and a secondary grid. The primary grid is based on the grid of the primary structure and the secondary structure of the façade is then aligned with the secondary grid. Hence the façade and bearing structure can be specified independently, and elements can be arranged at an offset. The recurrence of the geometrical relation between the shell of the building and the façade, i.e. the primary and the secondary structure, facilitates the alignment of joints and other details.

There are two basic grid types, whereby both can be combined in various ways (10).

Centreline grid: The base grid is aligned with the centreline of the building components. The length of the centreline is not defined. This can be particularly useful if the sizes of some or all components are not yet known.

Modular grid: A modular grid describes the extrapolation of the primary structure. The secondary grid of the façade is aligned with this primary grid. Zones with visibly varied widths are created in areas b and c.

Primary and secondary grids at an offset: Offsetting the façade grid in relation to the secondary grid can have an intermediary effect. However, this needs careful consideration when designing the wall joints. Sometimes intermediate members (c) have to be inserted for adjustment, or they can be used as an optional design element.

One of the most important attributes of a grid is that its definition entails a design decision. The organising principle of the grid is expressed in the façade. It defines the façade's proportion and rhythm. And choosing a particular grid helps to determine the horizontal and vertical arrangement of façade elements.

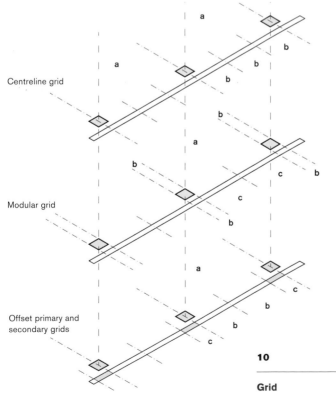

10

Grid
a) Centreline grid. b) Modular grid.
c) Offset primary and secondary grids.

Deciding on the position of the façade in relationship to the load-bearing structure of the building is one of the primary considerations in terms of the design and structure of the building (11-12).

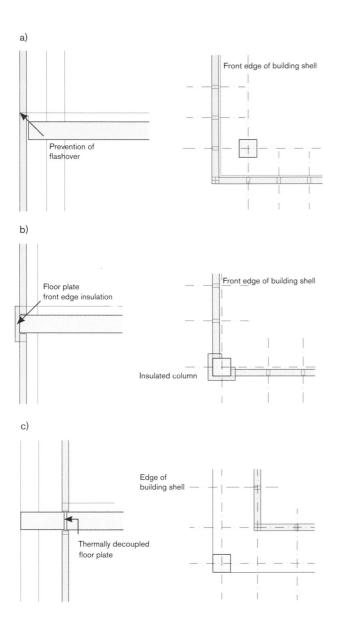

**Atlasgebouw Wageningen,
van den Oever, Zaaijer & Partners Architecten, 2006**
The façade lies behind the building's loadbearing structure.

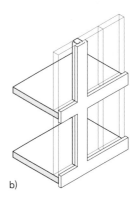

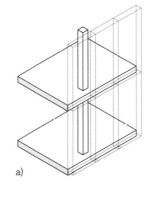

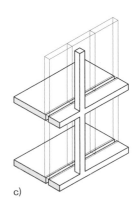

12

The façade's position

a) The secondary structure of the façade is positioned in front of the primary structure of the building. The shape of the ceiling slab in front of the column can vary. If the grids were aligned the size of the corner elements would be predetermined. The corner is transparent. It is possible to avoid showing the ceiling slab in the exterior façade grid. When doing so special consideration needs to be given to the space between the façade and the shell of the building for fire protection.

b) The façade is flushed with the primary structure. The surface of the ceiling slab needs to be insulated due to thermal requirements. The column's position creates an enclosed façade corner.

c) The façade is situated behind the primary structure. In this example the ceiling-floor unit penetrates the building's insulation, and therefore needs to be thermally decoupled. The column stands unattached in the outer corner.

Systems used in façade construction

On surveying current building trends, it becomes apparent that almost all buildings use systemised façades. This means that specific parts of the structure comprise standardised components provided by façade suppliers. So why do we need systemised solutions and how do they affect the planning and design of the façade?

Designing façades used to be part of the architect's job (13, 14). Previously, technical considerations such as thermal transmission coefficient or resistance to wind-driven rain were not relevant in the design process as the requirements were more basic. Problems arose when actual leakages or similar defects occurred. Much has changed since then, at least in the industrialised world. Because technical requirements have increased significantly, they are now fully regulated and can only be fulfilled by adopting sophisticated methods.

The necessity for systemising the façade is obvious, as the high demands of building performance now render the façade a particularly complex building component.

In addition to ease of design, systemised solutions offer the benefit to contractual parties of a predictable scope and sequence of the construction, from design and tender to the work in situ – resulting in better process control. This is also true for the dimensional tolerances allowed for a specific project.

Manufacturers test their systems for resistance to wind-driven rain, thermal insulation, air permeability, sound insulation, fire-resistance and building security. The design of the glass fixtures and the load-transfer joints between the post-and-beam sections are factory-certified. It is therefore possible to pick and choose from various systems.

The following applies to all systems: the actual task of design is to find applicable solutions for the system interface, i.e. the connection with other components. Within the individual system standards can be applied. However, for design and application, it is very important to know the strengths and weaknesses of each system under consideration.

It is the architect's job to specify the performance and technical requirements and to consider building regulations as well as those related to fire protection, sound insulation and thermal protection. In addition he/she is responsible for specifying the loadbearing structure, the façade and elemental grids as well as for determining the connection methods.

The actual construction planning at the construction site is then done by the contractors. The architect cannot possibly know all the details of the system. A portion of the production and assembly process is therefore beyond his control. This means that existing planning methods and communication processes during the execution of the construction need to be adapted to the new concepts and manufacturing methods. This is the only way to achieve efficient and safe process execution.

13

Crown Hall, Illinois Institute of Technology, Chicago, Ludwig Mies van der Rohe, 1956
The façade consists of a combination of steel sections. This design captivates us with its clean combination of materials, structural system and formal appearance.

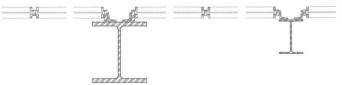

14

Crown Hall, Chicago, façade detail
This solution does not provide any thermal protection according to today's criteria. Resistance against rain penetration was considered during detailed design but a secondary drainage system, for example, is missing. The entire execution of the façade was done in situ.

Post-and-beam construction

Since post-and-beam is a widely used type of structure we want to provide a closer description. The base structure consists of loadbearing sections made of timber, steel or aluminium which assume the structural function of the façade (15). The sealing system on the interior is mounted onto this structure. Typically, aluminium sections are designed specifically to absorb the loads from this sealing system. Next would be the layer of the infill elements. These can be glass panes, windows or doors. The infill elements are mounted onto the posts and beams via mouldings that also constitute the outer sealing system. The load of the elements is transferred into the beams through support blocks. Water will always penetrate through the exterior sealing system into the construction. However it is channelled through the interior beam sealing into the interior post sealing system. At the base, the water must then be safely drained to the outside. The execution of the interface between beam and post sealing is therefore particularly important.

Depending on the glazing required or insulation value of the sections, different sealing systems can be used and combined. The design of the posts, beams and cover strips are generally independent of the system. For example, the system can also be mounted on a loadbearing timber post (17). The post's profile is designed specifically to accommodate the sealing system.

Typically the posts are mounted to the shell of the building with three-dimensional brackets (16). Then the beams are mounted, followed by the sealing system with the glass elements. The perimeter connections are next. After the mouldings are mounted, the structure is sealed. By default, the visible width of the sections is between 50 mm and 60 mm. Since the infill elements have to stay safely in place without slipping out of the seals when the façade deflects, this system does not really allow for narrower sections.

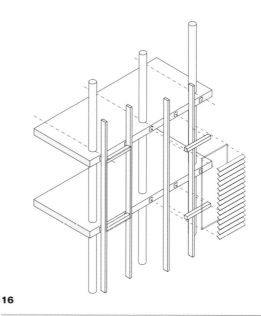

16

Assembly process of a post-and-beam structure
Typically the beams and posts are assembled in succession.

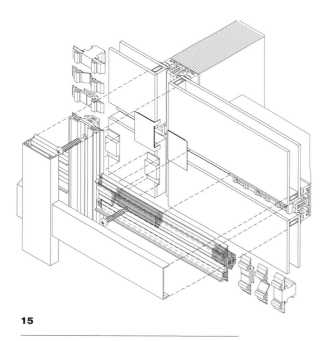

15

Post-and-beam construction
Perspective drawing of a junction.

17

Stacked timber structure for a post-and-beam façade
The secondary structure of the façade consists of loadbearing timber posts and beams. A sealing system is mounted onto the posts forming the interface to the exterior elements, e.g. the infill elements. Aluminium brackets are attached to the beams (as shown).

Unit system façade

The second most well-established façade system is the prefabricated unit system façade. The most significant difference between this type of façade and the post-and-beam façade is the degree of prefabrication. The goal is to reduce cost-intensive in situ assembly and man-hours, and to improve cost estimation. One of the major advantages is that manufacturing can be shifted to an earlier process phase and assembly can be carried out less dependent of the weather.

With unit system façades the glass elements as well as certain building services components can be pre-assembled to a great extent. The mounting parts on the shell of the building must be aligned accurately before the façade elements are installed. This is done storey by storey from the ground up. To avoid intersecting sealing sections the individual elements are equipped with a horizontal continuous sealing rail. Push-fit seals integrated into the elements are used to connect them laterally.

The linking of independent units results in double-splice profiles which increases the visible width of the sections. Therefore the standard post width is approximately 2 x 40 mm; twice the width of a single element. This means that the allowable degree of transparency is lower than with a post-and-beam façade. The goal is to produce elements as large as possible. Their size mainly depends on the transportation options. Typical dimensions are one storey high and 1.20-2.70 m wide. However, elements with a height of several storeys and a width of multiple modules can also be used.

Designing with systems

Using systemised solutions always implicates a constraint on creativity because the system product already provides a standardised solution by default.

Therefore architects try to exert influence on the system products to realise their designs. The demand for smaller components and higher transparency is developing in the production of systems for large-format glazing.

In most cases the creative design idea for a project is based on the perception of the building as a unique product. One reason for this is that complex building projects have very specific requirements. But it is also rooted in the architect's conception of him-/herself as the creator of a unique product. Special designs, however, stand in stark contrast to the ideal of systematised building. In some cases it is possible to realise a special design by adapting an existing system within its permissible limits. If these limits are too restrictive a new product has to be developed with all necessary tests and certifications. Both processes require a high degree of knowledge about the systems and close collaboration with the industry and manufacturers. Specialised façade planners are consulted for this process. Modification of a system by the architect can only be realised if the manufacturer can anticipate an increase in product market value that ensures a return on the investment. The result is that more often than not existing system solutions are used that approximate the architect's design. Typically, budget restraints prevent system adaptation or new system development. However, there are exceptions – mainly in major projects such as high-rise buildings. Here customised solutions may be of interest and economical because of the large number of units needed.

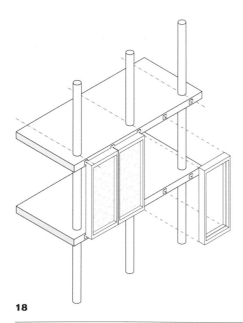

18

Assembly process of a panel system façade
Prefabricated elements being assembled in situ.

19

Assembly of window units in the factory
Prefabrication is one option to increase quality and quantity at the construction site.

Considering the fact that the majority of projects built in Europe is carried out with cast-in-situ concrete and detached systemised façades, the question arises about how much creative freedom there really is. The topic is revisited many times. The systemised façade concept already rules the thought process when designing a façade.

Openings in façade constructions

Openings are an important topic for all types of façades. Openings allow us to link the exterior and interior environments in a controlled manner. They provide interior and exterior views, ventilation, and regulate the transfer of humidity and sound. Openings can be of various sizes and serve multiple purposes: Entry and exit ways for people and vehicles, emergency exits; temporary inspection access, openings for cleaning, or technical installations, conduits (19).

The openings' orientation, location and dimension are closely linked to the purpose and usage of the interior space. The shape of the reveal, for example, has a major affect on the natural lighting of the interior space. The location of the openings can either facilitate or hinder natural ventilation. Adjustable openings are necessary to ensure a safe indoor climate. Fig. 20 shows standard solutions. The opening method can be achieved manually or automatically.

Hardware

Certain hardware is needed to allow for operable opening elements within the façade. Hardware fittings such as hinges or stays constitute the connection between element and façade construction and are used at the exterior interface between secondary structure and infill elements. Parts necessary to operate the elements such as door handles also fall in this category.

From a structural point of view the hardware elements need to be coordinated with the structure of the specific façade system in use. The loads of movable elements must be transferred into the secondary structure of the façade. Many system manufacturers therefore offer standard solutions. If motorised fittings are used (22), for automatic ventilation systems for example, the control of this system must be attuned with the entire electrotechnical concept of the building. Conduits also need to be planned for during façade construction. Hardware fittings feature significantly in the detailing process of the façade.

Façade requirements	Effect on the fitting
Type of opening	Type of fitting
Design	Type of material, shape, possibly concealed arrangement
Operation	Manual operation (handle, motorised system); position of controls
Opening clearance	Specification of tilt and turn fittings
Size and weight of element	Type of material, size of fitting
Frequency of use	Type of material, size of fitting
Safety features (building security, fire protection, emergency exits)	Appropriate safety fittings

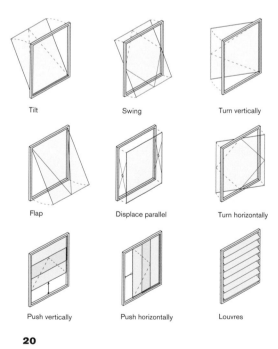

Tilt   Swing   Turn vertically

Flap   Displace parallel   Turn horizontally

Push vertically   Push horizontally   Louvres

20

Different types of openings
Window constructions with different modes of operation.

Windows

Windows are available in all kinds of materials. The choice of material has repercussions on the structure and the design of both window and façade. Therefore the most common window constructions and their specific materials are being introduced.

Timber windows and timber/ aluminium composite window

Timber window constructions are based on developments over a hundred years (21). As such, there is a wide variety of designs in use today. The type of timber used must demonstrate resistance to temperature and humidity fluctuations as well as pest-resistance. When designing timber windows (23) several factors need to be considered:

• Water penetrating into the construction must be carefully re-channeled to the outside.
• The rebate area inside the construction must be vented.
• Exposed edges need to be appropriately spaced in relation to other components so that they can dry out completely.
• The edges of the window frame must be chamfered carefully.
• Water should not be allowed to penetrate into the corners of the window. That is why timber windows typically do not have mitred joints in the lower corners. Instead the lateral frame section is one continuous part from top to bottom.

Timber windows have to be impregnated with wood preservative to protect against mold and insect infestation. The surfaces of timber windows need regular maintenance and periodic re-coating. The lower sections of the window frame are particularly exposed to the weather. Therefore metal weatherboards are often mounted onto the structure. One variant is the timber/aluminium window (25) with an aluminium cladding covering the entire exterior window frame. Well-maintained timber windows can last a very long time. The psychological aspect of timber plays a major role. Timber is nice and easy to work with and is conceived as a 'warm' material. However, ecological aspects need to be considered. The use of tropical wood, for example, is disputable in spite of its longevity if the origin is not certified; generally, wood from sustainable forestry is preferred.

21

Traditional window fitting
Fittings on folding shutters with a steel sleeve in the stone column as counterpart.

22

Motorised windows
Motorised windows with concealed conduit in the frame structure.

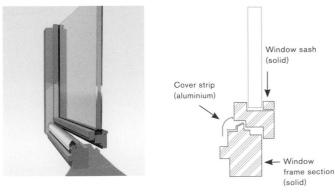

23

Schematic representation of a timber window
Timber window constructions are based on developments over hundreds of years. As such, there is a wide variety of designs in use today.

24

Extruded aluminium sections
The design of the mould permits very detailed profiles.

26

Corner bracket
The inserted corner bracket connects two aluminium sections. The edges
are then bonded and pressed.

Aluminium windows

Fig. 24 clearly shows the unique cross-section of extruded aluminium sections that are used for aluminium windows. The design of the mould permits very detailed profiles. Rubber seals
can be inserted directly and reinforcement bars provide structural integrity. Since aluminium is an excellent heat conductor
this type of window consists of an inner and an outer shell that
are connected by heat-insulating plastic profiles (27). The sections are therefore called aluminium-plastic composite sections.
The corner joints can be plastered and painted. The sections are
cut to length and necessary recesses for fitting components and
similar parts are milled prior to assembly. The sections are connected with corner brackets (26), glued, pressed or riveted in
place. In rare cases the sections can be joined by welding prior
to coating, then ground and painted.

Aluminium windows have several advantages: maintenance
is simple and undemanding. They are easy to work on and feature high manufacturing accuracy that translates to very close
tolerances and thus tightly sealed joints. In the long term these
properties can compensate for the higher purchase costs – one
reason why they are mainly used for large projects. Aluminium
windows offer good heat and sound insulation properties.

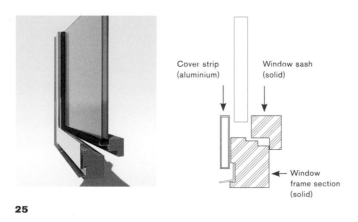

Cover strip
(aluminium)

Window sash
(solid)

← Window
frame section
(solid)

25

Schematic representation of a timber/aluminium window
The weathering side of the window is protected by aluminium cladding.

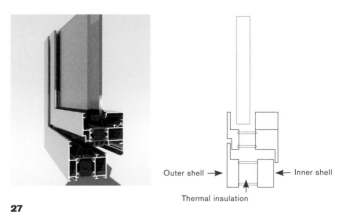

Outer shell →   ← Inner shell

Thermal insulation

27

Schematic representation of an aluminium window
Since aluminium is an excellent heat conductor this type of window consists
of an inner and an outer shell that are connected by heat-insulating plastic
profiles.

Steel windows

Steel windows are assembled from cold-rolled hollow sections. The profiles are made by folding the sheet metal. As with aluminium, steel sections require thermal separation of the inner and outer shell by means of heat-insulating plastic profiles (28). New window sections made of stainless steel reduce the amount of material in the centre zone, specific recesses decrease the heat transfer, and a plastic section as thermal separator is no longer necessary.

Steel sections are characterised by high bending and torsion strength. This might be advantageous, especially if the structural integrity of the frame is essential. However, they are more expensive than comparable aluminium options. Special care must be exercised regarding protection against corrosion. Here stainless steel sections offer particular safety in use. Steel sections come off well when comparing fire protection properties and are thus often the first choice.

uPVC windows

Similar to aluminium windows plastic windows consist of several sections. An extrusion process is used to manufacture these sections (29).

Numerous types of plastic materials are used for window constructions. However, the material most commonly used is uPVC, not least due to its impact and scratch resistance. But the thermal properties of uPVC are inferior to those of other materials used in façade construction. Solar radiation can cause dark sections to heat up to 80°C which in turn can lead to deformation. Coloured sections are made by adhesive bonding of a dyed top coat onto the base material. This process reduces the price advantage compared to other materials. Since uPVC windows are not particularly rigid, their installation sizes are limited. The sections are often reinforced with alloy tubing. The size of the fittings needs to be dimensioned accordingly.

The benefits of uPVC windows are easy handling; low cost and resilience during installation; protection against corrosion is not needed, eliminating any related problems. However, uPVC windows do not provide effective resistance against fire.

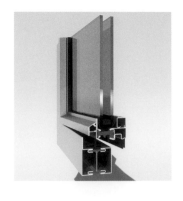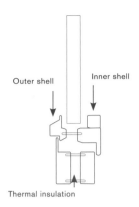

28

Schematic representation of a steel window
The folded steel profile is clearly visible.

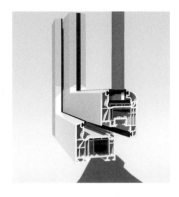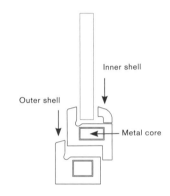

29

Schematic representation of a uPVC window
Metal rod inserts in the frame improve the structural integrity, realised here in the form of a metal core.

GRP windows

The application of glass-fibre reinforced plastic sections for window construction has been pushed on over the last years. GRP sections are manufactured by pultrusion, and due to the composite material's very good loadbearing capacity and low thermal conductivity they offer advantages over aluminium window sections. GRP window sections can be very slender and do not require insulation bars, but their application is restricted compared to aluminium sections due to significantly higher manufacturing cost. The manufacturing process is much more elaborate than an extrusion process, resulting in a very small product portfolio. From a constructional viewpoint GRP sections offer advantages and their slender profile appeals to architects. However, it remains to be seen how far the material will be applied in practise.

Assembly

The façade industry constantly searches for new manufacturing and assembly methods. The trend leans towards reducing in situ assembly times. This would shorten the construction period and reduce potential scheduling conflicts with other subcontract work. Assembly at the construction site also means higher risks due to changing weather conditions. At 5°C or less it becomes very difficult to install sealing systems safely. Assembly inside a factory building is usually cleaner and more controllable. Also, it is easier to resolve possible defects and problems.

However, prefabrication of elements into larger units does have its disadvantages: typically the elements are more complex and need to be dimensioned to account for stress during transportation. Immediately assembling the elements in situ requires extensive planning of the structural joints and permissible tolerances of the shell of the building are limited. Mounting units on the shell of the building must be installed with great accuracy because tightly planned construction logistics do not allow for delays. Transportation too can cause trouble. If materials (such as a glass pane) are damaged during transport this might not only mean the material itself needs to be replaced, but possibly entails re-manufacturing of the entire element.

All this raises several questions that need to be considered before choosing either the post-and-beam or unit system façade:

• What are the manufacturing processes of the contracting company and how knowledgeable is the team?
• What type of manufacturing equipment is available on the factory premises?
• What season will the assembly take place and how much time is allotted?
• What is the scope of the project and does it include a sufficient number of repetitive parts to warrant a systemised solution?
• What are the transportation options and what hauling devices will be used?

• What is the expected quality of the shell of the building in terms of tolerances?
• What are the properties and condition of the interfaces with adjacent subcontract works?

The construction of a façade is a process that begins with the architectural concept and ends with the assembly of the final product. However, this is not a linear process but rather depends on regular feedback (30) arising from complex decision making and communication processes. One example: while planning the assembly process it becomes apparent that a unit system façade solution is more economical then the type of structure previously chosen. This can lead to significant design modifications because the sectional width for this structure might be very different than that of the previous design. To achieve a well-controlled process it is therefore important that all members of the planning team possess a good basic knowledge of the principles of construction. Also, communicating all decisions made in the process is essential.

The construction industry is in transition from the traditional building trade to industrial production. Due to technological developments such as the Internet a vast amount of information about new materials and production methods is readily available – not only related to the construction industry but other disciplines as well. Architects and designers strive to put such new insights into practice. There is an enormous drive toward technical innovation and we can expect this to affect the construction of façades. Rising quality standards, shorter construction times and an acute awareness of energy consumption will help develop the façade into an increasingly complex product. Architects have to rise to the challenge and adopt systemised solutions as a viable design alternative.

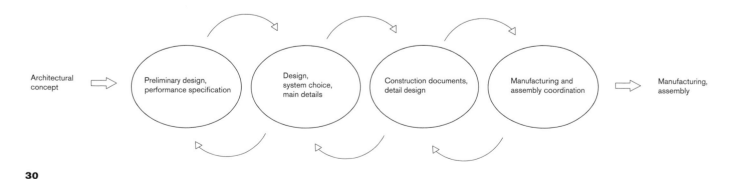

Architectural concept → Preliminary design, performance specification → Design, system choice, main details → Construction documents, detail design → Manufacturing and assembly coordination → Manufacturing, assembly

30

Design process for a façade
Designing a façade is not a linear process but rather depends on regular feedback.

4 | Detailing and Tolerances

Detailing is an integral part of the design process. The design process generates ideas for detailing solutions and ways of putting them into practice. Every detail is a key part of the design, and detailing problems reveal problems in design development. This may be illustrated with reference to the protection of traditional timber buildings. Projecting eaves play a key role in protecting timber structures in the façade from the elements, but also help to determine the character of these buildings (1). If they are omitted with the aim of achieving a clean modern profile, other effective means of protecting the timber elements of the building must be found. If this is not done, the building will weather rapidly.

It goes without saying that attention must be paid to details in order to achieve good design aesthetics, but proper detailing is also essential in the interests of structural integrity. It is impossible to construct a building that is attractive and stands up to wear and tear in the long run if detailing is neglected.

Apart from aesthetic design requirements, the detailing of modern buildings is made more difficult by the increasing complexity of the construction. For example, if details on the façade were used traditionally to keep rain out and to keep heat in (2, 4, 6), the same details nowadays are responsible for the functions of windproofing, protection against wind-driven rain, keeping the building cool in the summer and warm in the winter and preventing vapour diffusion (3). This increase in complexity reflects the separation of different functions in different layers of the façade and the use of proprietary systems for individual functions – e.g. multiple glazing systems, sealant systems and mullion systems (5). Detailing is thus reduced to systematic combination of the appropriate discrete components to perform the required functions against a backdrop of growing overall building complexity. While discrete elements may be changed in this process, the components used generally remain constant.

1

Traditional method of timber protection
Timber buildings will only last if consistent attention is paid to proper detailing of the structure. This example of a Swiss chalet clearly illustrates the role of projecting eaves in protecting the façade.

2

Window with timber frame and shutters
This picture shows single glazing and shutters in a half-timbered house. The shutters not only keep rain off the window but also provide a climatic buffer, allowing the house to retain more heat during cold nights.

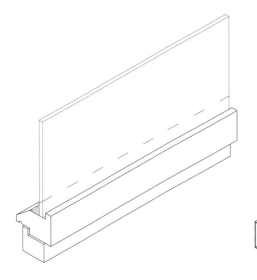

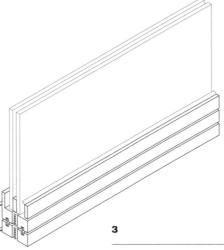

3

Traditional and modern window design
While traditional window design was limited to solving the problems of keeping rain out and heat in, modern window constructions have to meet more stringent requirements on protection against wind-driven rain and the effects of heating, thermal insulation, windproofing and prevention of vapour diffusion.

4

Traditional timber window
Ground-floor timber window. The timber sash of the casement shown in this picture closes against the metal window frame by means of an offset. A drip guard at the bottom of the frame protects against wind-driven rain. Plastic profiles for windproofing do not yet exist.

5

Modern timber window frame
Unlike traditional timber windows, modern timber windows are windproofed at the base by means of a number of folded seams and a silicone seal. In this example, a drip guard is again used to protect the sash against wind-driven rain. Water that does find its way into the frame can be led off via the drainage channel and an external aluminium weather strip.

6

Traditional sash window frame
Example of a traditional sash window. It is relatively easy to keep water out at the bottom: since the sash slides at the front of the frame, bevelling the sash at the bottom will suffice to exclude water. Sealing off the sides is more difficult.

Building grid and positioning of components

The position of the façade with reference to the rest of the building can be determined with the aid of a building-related grid. Buildings consist of surfaces, which are formed by combining individual elements or components. If apertures are created in these surfaces, these define a transition between one component and another. In order to organise the combination of these components and the joints between them, grids are generally imagined to be superimposed on the building to allow recurrent situations to be solved in a uniform manner. Such grids may also be applied to the component parts of the building – e.g. the masonry, in view of the constant dimensions of the constituent bricks (7).

The precise position of a component in a building can have important consequences. For example, an aperture can be closed with the aid of a recessed window. In this arrangement, the window is protected by the building, while from a visual point of view the front edge of the aperture will cast a shadow that tends to cut up the façade. A disadvantage of this arrangement is that if the thermal insulation properties of the wall are inadequate, a thermal bridge may be formed round the window.

Alternatively, the window can be placed as far out as possible to emphasise the uniform appearance of the façade or may even be projected in front of the façade. Here again, there is the potential disadvantage of a thermal bridge round the window; in addition, the windows are no longer protected in these arrangements. It follows that the best solution is probably to locate the window in an intermediate plane taking into consideration the resultant visual impression created on the façade (8).

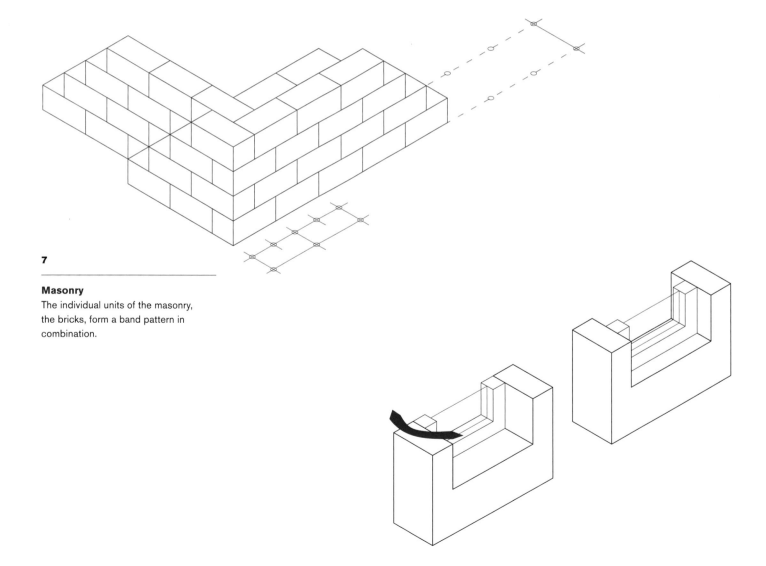

7

Masonry
The individual units of the masonry, the bricks, form a band pattern in combination.

Façades are influenced by different factors such as the façade's self-weight, acting vertically in the plane of the façade, and largely lateral wind forces acting perpendicular to this plane. If the façade is also used to carry the weight of the building as a whole, this load will also act as a vertical stress on the façade system (9). Other external factors acting on the façade include noise, wind, rain, heat, cold and solar radiation. The factors acting on the façade from the interior include air humidity, heat and cold.

In general, these various factors are considered separately in view of the different requirements they pose on the construction, and they are dealt with in separate functional layers in the façade.

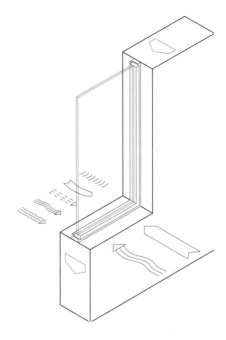

9

Factors influencing the façade
Façades are loaded in various ways by the overall structure of the building, the way in which it is used and the environmental conditions. External influences include noise, wind, rain, heat and cold, while factors acting from the interior include air humidity as well as heat and cold. The façade will also have to bear its self-weight and wind loads, and in certain cases loads derived from the structure of the building as a whole.

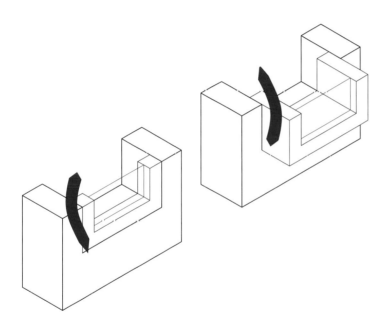

8

Position of window in building
The window may be placed in a number of different positions with respect to the structure of the building: recessed, in the median plane, flushed with the outer surface or projecting. The position in the median plane is the only one which can be detailed to avoid thermal bridges.

Combination of functions

When dealing with the appearance and the structural design of façades, we may divide them up in two different approaches, into functional elements (10) or into layered systems (11). In the former approach, each element will perform a particular task such as ventilation, lighting or the limitation of visual access. It is only when these elements are combined that the façade as a whole will perform all functions expected of it. Each element will perform its function at a particular part of the overall structure, and can in general be individually replaced if necessary (e.g. a dilapidated window unit can be replaced by a new one).

In a layered system, each function is performed by a different layer of the façade. The layers may be arranged so that the function can be performed at any point on the façade. Each function is performed via the layer in question to meet the relevant requirements. The complexity of the construction and the need to integrate the individual functions (such as ventilation, transparency and thermal insulation) are the challenges in this approach.

These two approaches are rarely used in isolation: many mixed forms and variants exist, and are designed to meet as many requirements as possible simultaneously thus giving optimum functionality with the most economic means. Continuing development is leading to the production of more and more specialised components, both in the field of façade elements and in that of layered façades.

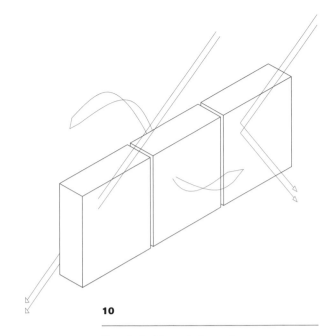

10

Façade built up of elements
This figure shows a façade composed of separate elements, each one of which performs a separate function such as ventilation, lighting or transparency control.

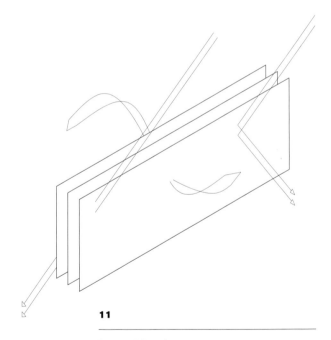

11

Layered façade
A layered façade has a uniform appearance but allows all desired functions to be realised at any point on the façade.

Detailing principles

Independent of the choice of materials and the desired appearance of the façade, two fundamental guidelines of façade design corresponding to the basic laws of building physics may be formulated here.

Firstly, water impinging on the building should be led off externally. If despite protective measures water does get into the building, it should be allowed to drain off or evaporate without harming the structure. This second proviso is necessary because no building can be guaranteed to be entirely weatherproof throughout its entire life cycle. If water does manage to penetrate the building fabric, this can cause timber to rot, steel to rust and (in the case of frost) masonry to fragment. To avoid this, a water drainage system covering the whole façade should be built behind the outer weatherproofing layer. If this is not possible (e.g. when sandwich systems are used), then some form of monitoring should be present and the façade components should be made water-resistant.

Secondly, the impermeability of the façade to air humidity should fall off from the inside to the outside of the façade. This means that water vapour from internal sources should not be able to penetrate into the building structure. If water does penetrate into the façade it will cause condensation when the external temperature is low, causing damage to the façade fabric. On the other hand, if water vapour from the exterior penetrates into the building fabric it should be eliminated through evaporation. Furthermore, an impermeable inner envelope avoids draughts, thus reducing heat loss.

Layering of details

In general, separate layers of the façade (each with one or a limited number of functions) are used to provide protection against different environmental factors (12).

An external weatherproofing layer offers protection against rain, wind and solar irradiation. As described above, a second drainage layer should be provided behind this. If there are windows in the façade, some additional means of reducing the impact of solar radiation before it reaches the windows can be provided in this layer.

An intermediate layer provides insulation against heat and cold in both directions. To prevent or minimise the direct transmission of heat, this layer should be thermally isolated from the outer weatherproofing layer. To this end, direct contact between the two layers at various points, allowing passage of heat or cold, should be avoided – though this is not entirely possible from a constructional point of view, since the intermediate layer supports the outer weatherproofing layer. In any case, the contact points should be kept as minimal as possible or heat flow should be minimised by use of materials of poor thermal conductivity.

If the intermediate layer is sufficiently solid, it can also assume the task of soundproofing. If it is not sufficiently solid, this layer must also be decoupled from the others to minimise the penetration of sound waves. If the façade also has a loadbearing function, this is performed by this layer too.

The innermost layer separates the interior space from the façade or from the external space. Windproofing and vapour barriers are localised here. In some cases, this layer may absorb water vapour from the interior space and return it to the interior space later. For the reasons indicated above, passage of water vapour through the impermeable layer should be avoided.

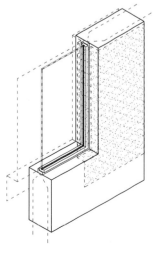

12

Layering in detail
Three layers in the façade detail provide protection against environmental factors. The outermost weatherproof layer keeps out wind-driven rain. The middle layer provides loadbearing capacity if required and also functions as thermal insulation. The innermost layer separates the interior from the exterior space and functions as vapour barrier.

The complexity of the connection of the three layers increases when structural solutions have to be sought for corners or façade-roof transitions (13). The desire to connect the layers without interruption round a corner gives rise to problems due to the different requirements imposed on the different structural components – vertical components (façades) have to be impervious to water, while horizontal components (roofs and eaves) have to be impervious to water and if necessary drain water off. Considerable construction detailing is required to take the layers round a mitred joint, since the decision about the overlapping occurs at the thinnest end where all layers meet. The problem is solved in a parapet construction by the use of specific profiles with sharp edges, but involves the difficulty of water exclusion from the roof on the inner side of the parapet wall. The use of eaves in this case deals with this problem by allowing the layers to project past the junction, but only at the expense of exposing the edges of some components which may compromise on aesthetics.

Examples of detail development

The principles underlying the development of details will now be illustrated with reference to five typical cases: masonry cladding, a post-and-beam façade, a unit system façade, a parapet and a plinth construction.

14

Example of masonry cladding taken from a housing development in Middelburg, the Netherlands
This double-skin construction consists of a concrete loadbearing layer with thermal insulation properties and a separate masonry weatherproof layer. The masonry is held in place by tie rods, which pass through the concrete layer. Following a procedure that is common in the Netherlands, the window frame is mounted before the masonry wall is built, which allows the latter to be adapted to suit the dimensions of the window. A waterproof membrane is mounted in the wall above the window, to allow water that has got into the wall to drain off again.

13

Principle of methods used at façade junctions
The treatment of façade layers at junctions reflects the problems associated with separation of function: vertical components are impervious to water, while horizontal ones also have to provide drainage facilities. This drawing shows on the left, absence of eaves, which is bound to lead to damage in the long run; then eaves in which only the outermost weatherproof layer projects; next a parapet, where separate drainage has to be provided; and on the right, a solution in which the intermediate functional layers are superimposed while the weatherproof layers overlap.

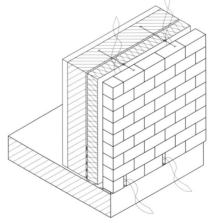

15

Sketch showing principle of masonry cladding
The loadbearing layer (here a concrete wall that also provides the thermal insulation) and the masonry weatherproofing layer are separated by an air cavity. Slits in the masonry act as air inlets. An impermeable membrane inserted into the wall is brought out near the bottom of the façade to allow water that got into the weatherproof layer to drain off.

Masonry cladding

One widely used double-skin construction is a masonry wall or a concrete wall made of prefabricated elements, which provides thermal insulation and loadbearing capacity, and an outer masonry skin for weatherproofing (14, 15). The thermal separation is provided by wire tie rods, which do penetrate the thermally insulating layer but hardly give rise to any heat flow because of their low cross-sectional area. These tie rods enable the outer masonry skin to resist lateral forces and the risk of bending due to its self-weight.

In this example, taken from a housing development in Middelburg, the Netherlands, a waterproof membrane passing through the double wall above the window allows water that managed to penetrate the façade above the window to be drained off before it reaches the window. The window frame was mounted before the masonry wall was built – quite a common practice in the Netherlands, which takes advantage of the prefabricated window elements and the possibility of adapting the faced masonry to fit the window-frame afterwards. The actual window with sash is mounted subsequently, to avoid damage during the building work.

The innermost layer (not visible in this example) is provided with a plaster finish, which gives not only complete impermeability but also thermal storage capacity.

Post-and-beam façade

Post-and-beam façades are made of storey-high posts, secured to the ceiling-floor units with the aid of mounting shoes, to which horizontal beams are connected (16, 17). This structure forms the loadbearing layer. The weatherproofing layer is formed by panels (sheets of glass or sandwich elements) fixed on to the post-and-beam structure with the aid of mounting frames. The loadbearing and weatherproofing layers are separated by spacers of low thermal conductivity and bolts fixed at appropriate points.

The separation is less clear at the layer of the glass sheets, since here the innermost layer, the loadbearing/thermal insulation layer and the weatherproofing layer are all combined in one.

Post-and-beam façades are made of prefabricated elements that are assembled by hand on site. This system works well, since the post-and-beam combinations are only mounted on the ceiling-floor units and are thus largely independent of the fabric of the building. As a result, tolerances can easily be corrected. A disadvantage is the necessity to close the gap between the ceiling-floor unit and the façade subsequently, to meet noise-control and fire safety regulations.

16

Post-and-beam façade:
Fachhochschule Detmold, Werkstatt Emilie,
2007
This picture of a post-and-beam system shows how the façade structure rests on the ceiling-floor unit. The mounting shoes are clearly visible. Glazing and cladding panels are fixed in place in subsequent stages, with the aid of mounting strips.

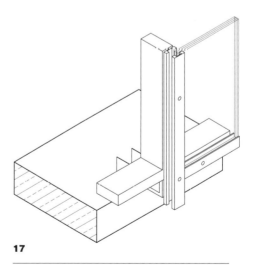

17

Sketch showing principle
of post-and-beam façade
The post-and-beam façade consists of storey-high posts to which horizontal beams are connected. The glazing and cladding panels are secured from outside with the aid of mounting strips, and are thus thermally isolated from the main structure.

Unit system façade

Unlike a post-and-beam façade, a unit system façade consists of fully prefabricated elements that simply have to be positioned and mounted in situ (18). This is also generally made storey-high, and usually consists of a loadbearing framework in which glazing and cladding panels can be infilled (19). Since each element is a complete unit, a given component of the system façade will usually perform several functions simultaneously: for example, as in the case of a post-and-beam façade a pane of glass inserted into the structure will not only separate inside from outside but also provide thermal insulation and weatherproofing.

Each façade element is connected to the building structure by means of an angled cleat mounted frontally near the ceiling. The element is suspended from the top, and stabilised against lateral forces at the base by a sliding bolt connection to the element below it. Successive elements can be combined in this way, either being stacked from bottom to top or connected in a row.

Since the individual modules are prefabricated and transported to the building site, they have to be fairly rigid. This can give rise to problems when sealing the gaps in and between the elements. Unlike the case of post-and-beam systems, where the mounting strips provide the seals in the weatherproofing layer, the sheets of glazing in unit system façades have to be individually sealed. In addition, an effective seal has to be provided between adjacent façade elements. This is usually done using three sealant profiles which have to be introduced in special grooves between the elements during assembly.

18

**System façade: Double façade
of Debitel Headquarters, Stuttgart,
RKW Architektur + Städtebau, 2002**
This system façade is built with a double façade solution. The individual elements are lifted into place with a crane, and require only a small labour force for assembly. They already contain all necessary components, so that further finishing (e.g. glazing) is not required.

19

Sketch showing principle of system façade
Unlike post-and-beam systems, unit system façades are built up out of fully prefabricated elements. Sealing strips have to be placed between adjacent elements to ensure complete tightness against wind-driven rain and wind from the exterior and water vapour from the interior.

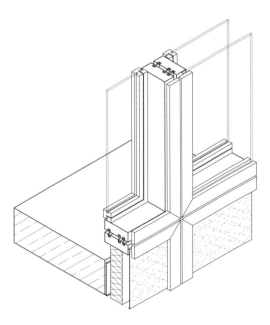

Parapet

The construction of a parapet on top of a building involves the problem of bringing about a suitable transition between the various layers of the façade and the various layers of the roof covering. The loadbearing and thermal insulation layers of the façade and the roof terminate at the base of the parapet (20), while the weatherproofing layers are continued up to the top of the parapet (the horizontal layers of membrane that form the roof covering have to be bent through 90 degrees for this purpose) – in any case far enough to prevent penetration of wind-driven rain or standing water on the roof.

To finish off this simple solution, a timber cap lined with membrane is placed on top of the parapet so as to cover the ends of both the façade layers and the roof-covering layers (21). An aluminium section with a drip guard is also provided at the front edge. An alternative – and possibly more durable – solution is to use a cap made of metal sheeting, which should also cover all layers involved.

20

Parapet on residential housing
In this case we see how the functional layers combining loadbearing capacity and thermal insulation come together while the weatherproofing layers are continued upwards. The membrane-lined timber cap used here could be replaced by a metal sheeting cap.

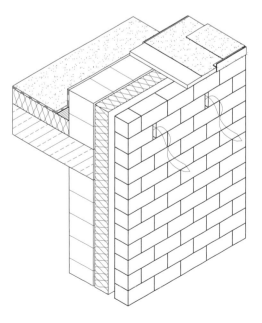

21

Sketch showing principle of parapet construction
The parapet is where the various functional layers of the façade and the roof come together. A suitable construction detail, providing adequate long-term protection against climatic influences, has to be devised here. The weatherproofing layers have to be continued vertically up to the top of the parapet on both sides, to prevent penetration of water. At this point, they are covered with a cap of membrane-lined wood or metal sheeting.

Plinth unit

The main problem in the design of a plinth unit is the transition from the façade to the foundations. Loads from the façade and the rest of the building have to be effectively transferred to the ground. In addition, a proper transition has to be realised between the weatherproofing of the façade and the foundation/soil vapour barrier.

For example, this detail could consist of a single-layer wall made of sheet-metal sections resting on a prefabricated concrete plinth (23). The layer providing separation from the interior space would in this case be made of recessed sheet-metal panels resting on the bearing structure of the building as a whole (22). The functional layer is the thermal insulation. The weatherproofing layer is made of sheet-metal panels on a support. Once again, the intermediate space is used to allow any water that may have penetrated this far to evaporate.

The prefabricated plinth consists of two concrete plates separated by a layer of high-density foam with sealed pores. This structure has the dual function of avoiding a thermal bridge and giving the plinth good impact strength.

This example is a good illustration of compliance with the above-mentioned requirement that the impermeability of the façade to water vapour should be graduated from inside to outside. The outer layer repels rain falling on it, but air can pass through the seams between the different elements and the gaps in the plinth unit. All openings are provided with drip guards to hinder the penetration of water. The thermal insulation behind this can dry out if necessary. The interior space is provided with maximum windproofing by sheet-metal cladding – which has to have suitable sealant sections in the gaps between the individual sheets.

It should be noted that there are of course other façade construction methods based on different principles, which may not always follow the principle of separation of the different layers described above. These other methods may be perfectly appropriate in given situations. The examples given above have been presented solely to illustrate the general principles of layering in building structures and how the transition between these layers is handled at junctions and points of penetration.

23

Plinth design
This cross-sectional view of the plinth detail shows the sequence of the functional and weatherproofing layers. The innermost layer is here formed by the sheet-metal surface. Sealing strips are used in the gaps between the innermost sheet-metal elements to ensure water tightness.

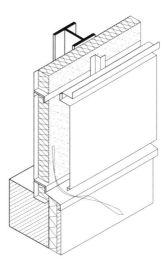

22

Sketch showing principle of plinth unit
The wall consists of a loadbearing layer of interlocking sheet-metal panels, within which the thermal insulation is introduced. A gap is left between the weatherproofing layer and the wall.

Joints

In order to realise the various functions required in the façade, it is necessary to combine a number of larger and smaller façade elements. These elements are generally connected by means of seams or joints between them (24). Care must be taken in this context to ensure that the joints do not interfere with the continuity of the individual layers or with the appearance of the façade as a whole.

A number of different types of joints may be distinguished visually, including hairline joints (gaps where the distance between the individual components is very small, though capillary effects can still lead to the risk of penetration of water), covered joints (25), joints with ridge reinforcement and false seams or shadow gaps (which look like seams on the surface, but do not correspond to any discontinuity in the underlying structure).

25

Joints in traditional timber window construction
Example of a traditional timber window construction showing different types of joints: the gap between the sashes is closed by a cover strip. The joints between the different components of the frame are realised as glued hairline seams. The bottom panels are connected by tongue and groove joints, while the panes of glass at the top are held in place with sealing putty applied on the exterior.

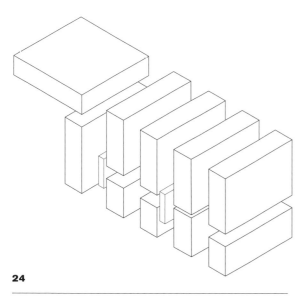

24

Types of joints
From left to right: covered joint, shadow gap, cover strip, hairline, open joint.

Apart from the effect of seams in the façade on the appearance of a building (26-33), it is also important to ensure that such seams do not impair the ability of the structure to exclude water, in particular wind-driven rain. To this end, open joints are covered or provided with a drip guard so that the water drips off harmlessly rather than penetrating into the building fabric. If joints are to be sealed, care must be taken in choosing the right type of sealant. If windproofing is also required, this is generally provided by another layer on the inside of the façade.

28

Slate covering

A slate covering is effective at keeping wind-driven rain out of the building fabric, despite the gaps between neighbouring slates, because of the overlapping manner in which the slates are laid. The top slate is always laid over the one below it with a certain overlap and then fixed on to the wall.

26

Larch shingle covering

Thin strips of larch can be used to give a covering similar in structure to that obtained using slates, which is both effective as weatherproofing and attractive. Since no two shingles are completely the same shape and size, a pleasing irregularity can be achieved with this form of covering.

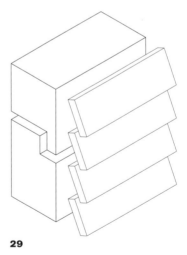

29

Open joint

To ensure good weatherproofing in the presence of open joints, projections that keep the rain away from the join or drip guards (projecting members with a specially curved edge to ensure that water drips off rather than flowing into the crack) must be used.

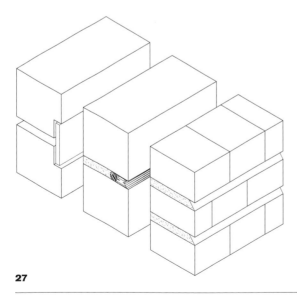

27

Sealed joint

Joints may be sealed in various ways to ensure good weatherproofing of the façade. Water can then only penetrate into the join if the sealant is damaged. This figure shows from left to right a cover strip with sealant section, a silicone seal and a mortar-filled join in brickwork.

30

Use of cover strips in pitched roof construction
When plank sheathing is used in the construction of a pitched roof, a cover strip is screwed over each join between successive planks so that even if the planks warp no open joints are produced.

31

Cover strip in a post-and-beam structure
In this case, which is similar to that shown in fig. 30, the cover strip used in a post-and-beam system covers the gap between two panes of glass. Additional silicone sealant sections provide a good seal between the glass and the cover strip. A problem with this construction is that appreciable stresses can be generated at the point – clearly visible in this figure – where the horizontal and vertical cover strips meet.

32

Silicone seal
Unlike the cover strip, the silicone seal does not involve forces between horizontal and vertical members, since the seal is produced by the curing of the sealant compound introduced into the gap, and the resultant seal has a certain elasticity. Disadvantages are that care and skill are needed to produce a first-class seal, and that the seal has a limited lifespan.

33

Masonry detail with sheet-metal cover
In this detail, the gaps between some bricks are filled with mortar while a sheet-metal drip guard placed some distance in front of the masonry keeps rainwater away from the open gaps left for ventilation purposes.

A closer look at the finished join will reveal the following picture, starting from the exterior: rainwater is kept out by an external sealing plane. Back-up protection against the entry of rainwater and drainage of water that managed to get in are provided by a second sealing plane. The innermost layer provides airtight separation of the interior space. Hence, even in the detailing of joints the individual functions are represented by separate recognisable layers (35).

Another function of joints is to take up movements of individual building components, many of which are predictable on the basis of design calculations. Where appreciable displacements are possible, joints must be capable of handling them. The sealant sections or sealant compounds used to fill the joints must be elastic enough to enable such movements (34, 36).

34

Concrete façade
In this view of a projecting concrete façade, the vertical joints between the different concrete elements are clearly visible. In this case, they are filled with a long-life silicone sealant. Also visible at the base of the façade are the drip guards used to allow rainwater to fall harmlessly off the surface of the building.

36

Joint between façade and roof, Chek Lap Kok Airport, Hong Kong, Foster and Partners, 1998
Since the roof of the airport building shows appreciable movement with respect to the façade, the join between them must permit a great deal of tolerance. This was achieved with the aid of a concertina-like plastic profile, which provides the necessary weatherproofing and windproofing. The requirements for thermal insulation are generally not very stringent at such locations.

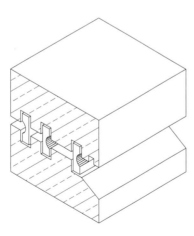

35

Construction of the seal in the gap between prefabricated elements
The seal in the gap between prefabricated elements shows a layered construction. The innermost seal provides windproofing, and the outermost seal provides weatherproofing. The intermediate seal provides a backup in case the weatherproofing fails.

Tolerances

The question of tolerances is an integral part of detailing. In the building industry, the term 'tolerance' is understood to mean the difference between the actual position of a given component with reference to the building as a whole and the position predicted on the basis of design calculations. As a result of such differences, the gap between some building components could become too large, while other components could be forced against one another leading to substantial stresses in the building fabric. Measures must be taken to ensure that these two extreme situations do not occur, so that the various parts of a building do actually fit together as they should and the construction process can proceed smoothly. In other words, the detailing of a building has to take into consideration not only the functions to be performed by the building and the (external) factors acting on it, but also the planned and unplanned changes in building dimensions. It may be noted in this connection that reinforced concrete elements generally have tolerances of up to 3 cm, depending on the size of the element (37). The tolerance for timber construction details may be taken to be in the range from 0.5 to 2 cm, and that for steel constructional details in the range from 0.2 to 0.5 cm. These have two consequences: in the first place, care must be taken to ensure that building components made of a certain material do comply with the tolerances specified for that material; and when different materials are joined together, the differences in tolerance between the materials must be handled correctly.

To solve this problem, connection devices must be designed to permit dimensional adjustment in one, two or all three directions. This may be done e.g. with the aid of slots in the connection device that allow the position of the mounting bolts to be adjusted over a significant length in the given direction (38, 41). Alternatively, bearing bolts may be designed so that their length varies with the load to which they are subjected, thus permitting motion of the component supported (39, 40). The most expensive solution would doubtlessly be the ad hoc positioning of the component in situ followed by fixing in the desired position (e.g. by welding).

38

Angled cleat with slots
Example of a slotted connection device, used here for the mounting of a pane of glass. The illustration shows the slots in various directions, used to adjust the position of the individual panes of glass, and drilled holes in situ to increase flexibility.

37

Tolerances in prefabricated concrete parts
Tolerances of up to 3 cm may be expected in concrete structures, whether cast in situ or prefabricated. These must be accommodated with the aid of specially designed connection devices, to ensure compatibility between the wide tolerance of concrete and the much more limited tolerances of steel and aluminium.

It is important to note in this regard that uneven joints caused by a failure to take differences in tolerance into account during the planning stage generally have an adverse effect on the appearance of a building. The extent of this problem can be delimited by making the joints wide enough or by concealing them.

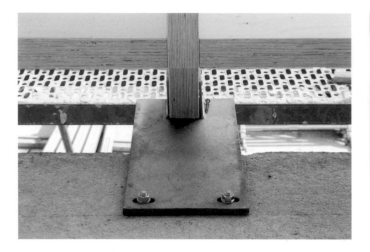

39

Point support for post-and-beam façade from inside
This point support for a post-and-beam façade is positioned in situ so that the axis parallel to the façade is kept flexible enough with the aid of a pin drilled into the concrete ceiling to permit fine tuning.

40

Point support for post-and-beam façade from outside
This view of the same support point shows that the axis perpendicular to the façade is positioned with the aid of a hole drilled in situ in the wood. If the façade were made of aluminium or steel, a horizontal slot would be provided to permit finer adjustment.

41

Point support for post-and-beam façade from above
In this final picture of the series, the top support point for the post-and-beam façade may be seen. A vertical slot is provided here, which may be used to accommodate both lateral tolerances and movements of the building structure itself.

Summing up, it may be stated that tolerances of the order of centimetres may be expected in reinforced concrete and timber building components, and of the order of millimetres in steel and aluminium. These tolerances must be taken into account in the detailing of buildings, in particular of façades where the commonly used combination of reinforced concrete and steel or aluminium may lead to problems related to differences not only in tolerance but also in thermal expansion between the various materials. Measures must therefore be taken to permit the adjustment of the connection elements to deal with the dimensional shifts that occur (42, 43).

42

Post Tower, Bonn, Helmut Jahn, 2003
Top view of façade base with connection points for fitting out like raised floors, mullions and partition wall.

43

Façade at base of Post Tower, Bonn
The unit system façade of the Post Tower in Bonn, designed by Helmut Jahn, is secured to the ceiling-floor unit with the aid of the mounting shoe that may be seen here. The picture also shows the means provided to allow adjustment of the mounting to accommodate differences in tolerances in all three directions: horizontally outwards with the aid of graduated slots and screws in the concrete surface, horizontally parallel to the façade with the aid of a locating bolt and vertically with the aid of locating grips provided with screw adjustment.

5 | Climate and Energy

Façade as interface to the exterior

The façade serves as the interface between the interior and the exterior space. Air and heat can be gained through the façade, but they can be dissipated as well. In order to provide the user with a comfortable environment, a façade must fulfil many functions. If the façade cannot meet the functional requirements by itself, additional components must be added in the façade layer or in its vicinity.

Functional requirements

The façade and the technical components interact with each other. The better the façade's thermal insulation is, the smaller the necessary heating elements have to be. And the more efficient the sun protection is, the smaller the necessary cooling units have to be. In some cases active cooling can be completely avoided, depending on the climatic conditions and interior heat loads. The façade is one of the most significant contributors to the energy budget as well as the comfort parameters of a building.

The example (1) demonstrates the impact that the quality of the façade has on the energy demand of a specific building (typical air-conditioned office building in Central Europe). It shows that the energy consumption in the interior space might decrease depending on the quality of the façade (heat and sun protection). It is clearly visible that a larger glass surface area offers a more economical operation; in this case, however, exterior sun protection is mandatory. The energy demand as well as the optimum glass area can, of course, be reduced or modified by employing additional passive and active measures.

The following sections describe passive measures (façade) and active measures (technical components) and their influence on user comfort.

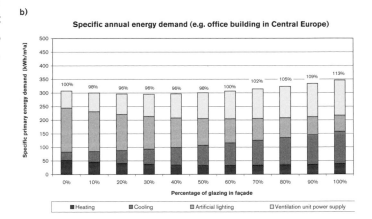

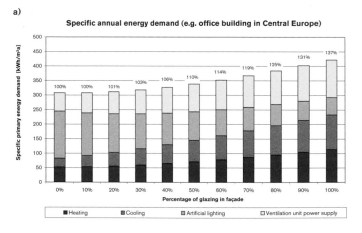

1

Primary energy demand of an administrative building
Specific primary energy demand of an administrative building in a moderate climate dependant on the percentage of glazing surface and the quality of the heat or sun protection of the façade. Diagram a) shows the energy demand with state-of-the-art double glazing and internal sun protection. In diagram b) the heat insulation glazing has been replaced by a triple glazing. External sun protection was added in diagram c).

User comfort

Different types of buildings such as residential housing or office buildings pose different demands on the comfort level. The most essential criteria are thermal, hygienic, acoustic as well as visual comfort.

All participating consultants should agree upon the many parameters that need to be considered during design. Examining particular aspects in isolation from the others might compromise the other requirements (2).

Each user defines comfort differently; therefore comfort levels cannot be measured with an objective measuring method for all users alike. When specifying comfort-related factors such as air movement, temperature, light intensity and humidity we can only aim to provide recommendations based on guideline values. We have to assume that each user perceives these differently and therefore feels more or less comfortable in any given environment.

Minimum requirements related to work environment or living space conditions are regulated by law, but in most cases these laws only serve to ensure the most basic criteria. Special comfort-related requirements should be specified by the participating consultant team members. The following section explains each comfort factor in more detail.

Thermal requirements

The human body not only absorbs and emits heat through the air by convection, e.g. transfer of energy through tiny particles in the airflow, but is also influenced by the surrounding surfaces through radiation. Therefore heat transfer by both convection and radiation needs to be considered when trying to achieve thermal comfort.

Because of these heat transfer mechanisms, temperature is specified as 'felt temperature' or 'operational temperature'. This measurement, also known as room temperature, corresponds approximately with the mean value of the air temperature in the room and the mean radiation temperature from the enclosing surface areas of the room. This shows how much impact the surface areas of a space can have on thermal comfort (3).

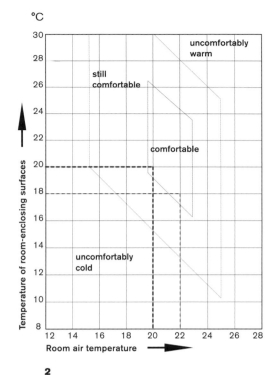

2

Comfort
Comfort range depending on room air temperature and the surface temperature of the room-enclosing surfaces.

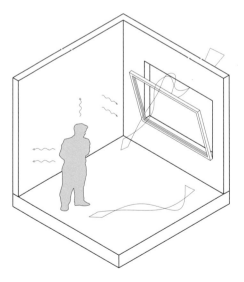

3

Parameters influencing thermal comfort
Many factors are responsible for the thermal comfort level. The human body emits heat through radiation and convection, but also perceives the heat/cold from the surrounding walls and the airflow in the room.

The specifications of mandatory temperatures or temperature ranges for rooms and buildings are regulated by many legislative directives of the individual countries. Generally, temperatures should always be evaluated in relation to the outside temperature. A difference of 5-6 K (temperature differences are specified in Kelvin, with 1K equalling 1°C) compared to the outside temperature has proven to be a viable definition whereby room temperatures of more than 26°C should be avoided.

Research has shown that users show higher acceptance of the room temperature if the temperature can be regulated by operable windows. Users are typically less satisfied if the temperature is controlled by a central air-conditioning unit that they cannot regulate individually.

Evaluation of the comfort level

The method of calculating the comfort level according to DIN EN ISO 7730 (DIN = German Industry Norm) enables consultants to estimate the user comfort level depending on the room temperature, the type of activity performed and type of clothing worn. This exemplary method of calculation provides a predicted mean user rating (4), from which a predicted percentage of dissatisfied users can be derived. (PPD = Predicted Percentage of Dissatisfied). The method is based on the thermal balance of the human body with clothing and activity level (5) as influencing factors as well as air temperature, mean radiation temperature, relative air flow and humidity. The goal is to strive for a percentage of dissatisfied users lower than 10%. This guideline regards the user as an individual with his or her unique sensations, and correspondingly, unique comfort level.

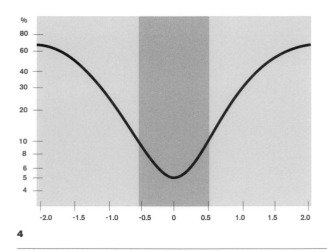

4

DIN EN ISO 7730 Calculation of the comfort level
Predicted percentage of the dissatisfied (PPD) as a function of the predicted mean vote (PMV)

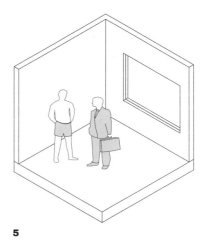

5

Clothing insulation values
German Industry Norm DIN EN ISO 7730 assigns a value to individual types of clothing. This makes sense considering that for bank personnel, for example, business attire is the proper dress code whereas employees of a marketing agency, for example, might dress more casually conforming to concurrent weather conditions.

Visual requirements

The goal related to the visual perception of a room is to please the eye of the occupant. As with thermal comfort, the users' visual perception as well as their preferences can differ significantly. In general, rooms should be designed such that the human eye can easily grasp the surroundings and receive a clear impression of the space. Easy navigation, sufficient lighting and minor differences in contrast facilitate the perception within a room and promote visual comfort (6).

Current research is underway to examine whether well thought-out design can compensate for inadequat environments. The effect of colours in a room should not be neglected either.

And another aspect of visual comfort, just as important but often underestimated, is natural light. In as far as incident sunlight is available it should be used. Human metabolism requires sunlight. However, to avoid overheating of the room and glare at the work place, sun protection is necessary. A stark contrast between light and dark areas resulting from cast shadows is also problematic. Thus, we have to compromise when planning for thermal comfort.

6

Visual comfort
Glare, reflections and stark contrast between dark and light areas such as the stripe-effect caused by sun sun shielding cast onto the concrete columns can reduce the visual comfort.

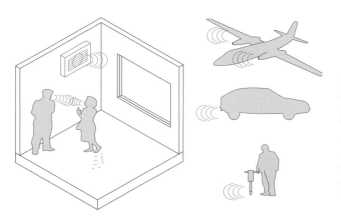

Hygienic requirements

A comparative research study of air-conditioned offices and naturally ventilated rooms conducted by the BMFT (German Federal Ministry of Research and Technology) in 1998 has found that the occupants of air-conditioned rooms felt uncomfortable more often than those in naturally ventilated and lit rooms. Working in air-conditioned spaces therefore influences the productivity level of the employees as well as that of the company. Fatigue and poor concentration resulting from such conditions are called sick building syndrome (SBS).

The quality of the ambient air plays a significant role in terms of hygienic comfort. Besides many other factors it is determined by the quality of the air supplied from the outside on the one hand and by the degree of contamination contingent on the user and the room furnishings on the other. Dust, gases, CO_2, odour substances, viruses and bacteria constitute such contaminations.

To ensure hygienic comfort the air has to be adequately circulated. We understand this phenomenon when we think of entering a holiday flat that hasn't been used for some time. The first thing we do is to open the windows and doors because it smells stuffy and muggy.

Acoustic requirements

The acoustic comfort level in a room is influenced by sounds transferred from the outside, sounds inside the building and from the person's own sound generation, or rather, the resonant response (7).

Noise from traffic and construction sites is the biggest source of sound from outside the building. Within the building one source of noise is the user himself or herself: talking on the telephone, walking around or listening to music. We have to differentiate between air-borne sound that spreads from the source through the room through the air and structure-borne sound, which spreads through the building components such as footfall sounds from heels clicking on hard floors.

Noises can also be caused by technical installations and conductors. Such sounds can spread through the whole of the building and therefore reduce the acoustic comfort level.

7

Acoustic influence
The acoustic influences that can affect a room comprise of exterior sound sources such as aircraft noise, and more often traffic and construction noise. In addition there are sound sources from the inside of the building that reduce acoustic comfort level, such as conversations in neighbouring offices, machine-operating noises as well as footfall sounds due to insufficient sound insulation.

When examining the building requirements individually we realise that they can be contradictory. For example, from an acoustic point of view, it might be necessary to suspend the entire ceiling with acoustically effective ceiling panels. Cladding the rough concrete of the ceiling, however, reduces its thermal storage capacity and therefore the concrete mass's natural cooling effect during summer. If suspending the ceiling cannot be avoided and the room's acoustics cannot be improved by using partitions or sound-absorbing furnishings we need to consider other methods for cooling the space. This example shows that the consultant team has to keep monitoring all demands on the entire building and accommodate them.

If the façade needs to provide a higher degree of sound insulation due to external noise, but operable windows are planned to offer natural ventilation, these would no longer ensure sound insulation when open. An alternative ventilation method has to be considered or we need to modify the façade such that the sound insulation is acceptable even when using the windows for ventilation.

Regulating the comfort level with the façade

The functions of ventilation, heating, cooling, sun protection and directing of light have to be realised through elements of the façade or by means of building services components in order to achieve the required comfort levels described previously (8).

Ventilation

As we have already seen in the previous discussion about comfort, ventilation is a vital factor. The users themselves strongly influence the environment of the rooms they occupy by their mere presence. The human body releases several litres of water per day into the room atmosphere in the form of vapour, depending on the type of activity performed. Exhaling raises the CO_2 content and the temperature increases. The CO_2 level should be kept at a maximum of 0.1-0.15%. Ventilation regulates the temperature as well as the relative humidity of a room. Exhaust air is replaced with fresh air and harmful as well as odourous substances are removed. Natural ventilation is regulated by respective norms and guidelines. There are two different methods of ventilating a room: natural and mechanical ventilation.

Natural ventilation

Natural ventilation includes gap ventilation, window ventilation and shaft ventilation (9).

Gap ventilation: Self or gap ventilation is the exchange of air in a room occurring when windows, exterior doors and roller shutter housings are closed but air penetrates through their joints due to the drop of pressure between the interior and the exterior, caused by temperature differences and wind incidence. Modern windows typically no longer permit gap ventilation since they are well sealed but some models comprise small operable flaps (10-12).

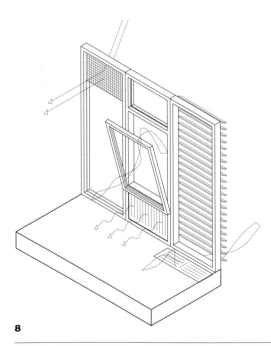

8

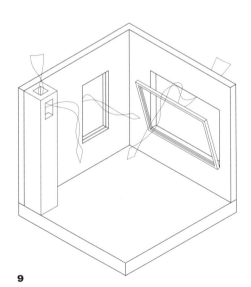

9

Overview of façade functions
The functions of ventilation, heating, cooling, sun protection and directing of light have to be realised through various components in the façade or in close proximity to it.

Natural ventilation
Natural ventilation can be divided into three categories: 1) gap ventilation with air being supplied through leaks in the frame or through dedicated small ventilation flaps; 2) traditional window ventilation, and 3) shaft ventilation where the exhaust air is drawn out through a vertical shaft.

10

Closed window with open flap
Adjustable gap ventilation in a wooden window; the arrow points to the open flap that provides ventilation even when the window is closed.

11

Closed window with closed gap vent
Closing the ventilation flap makes the window fully windproof.

12

Ventilation slots
Small ventilation slots with fly screens are mounted on the exterior of the window frame. They provide ventilation even when the window is closed.

Window ventilation: The most common method of natural ventilation is window ventilation whereby different types of hardware and fittings have an impact on the efficiency of the ventilation (13). Ventilation efficiency is also influenced by the wind pressure exerted on the façade. If ventilation is only provided through one side of the façade, it can be achieved for rooms about 2.5 times deeper than high. Cross ventilation causing 'draught' is more efficient, since sufficient ventilation can be achieved for rooms 5 times as deep as they are high. Today, motorised windows are available that provide automatic ventilation depending on the actual requirement or to facilitate opening windows that are difficult to access.

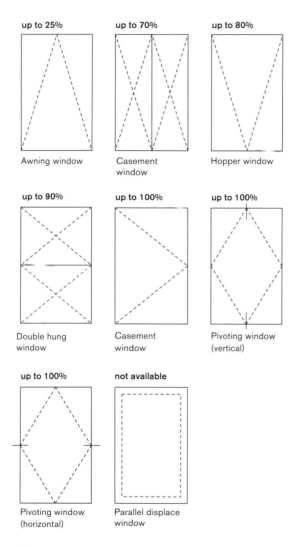

13

Different types of ventilation openings
The method of opening a window can be varied by using different hardware fittings. Each method entails a different amount of air passage.

Shaft ventilation: Shaft ventilation is primarily used to exchange high volumes of air. Fresh air flows into the room through the windows, is then exhausted through a shaft, usually located in the centre of the building, and exits through the roof. Shaft ventilation is widely used in apartment buildings and, in earlier times, was often used to allow large building depths. Shaft ventilation is also very reliable in winter because the shaft's wind-protected location within the building prevents the exhaust air from cooling rapidly and therefore maintains the ventilation's functionality.

Nowadays, exhaust shafts are often situated within the façade: the shaft-box façade is one example. With this type of façade the exhaust air is heated by solar radiation and therefore rises more quickly.

Mechanical ventilation

Mechanical ventilation systems are often employed if continuous ventilation is needed throughout the year. The required air exchange rates are regulated by local law.

The most simple mechanical ventilation systems comprise small motorised fans that are installed in the exterior wall (14) and blow the exhaust air to the outside. Fresh air is introduced from a different location.

A large quantity of building services are required for centrally controlled ventilation and air-conditioning systems because supply and exhaust ducts have to be run through the floor-ceiling units (15). The ventilation control stations take up large amounts of useable space; in high-rise buildings even entire storeys are designated as service storeys that contain all building services systems in one area. The big advantage of mechanical ventilation systems is the possibility of waste heat recovery: energy is drawn from the exhaust air and is re-introduced into the heating system. In many cases the air is conditioned as soon as it is centrally processed; this is then called an air-conditioning system. Particularly with regards to required space and individual control of the room environment, the trend goes toward decentralised units that permit individual regulation of the air-conditioning or ventilation of specific rooms or a defined group of offices (16). Such decentralised units consist of compact subsurface devices that are installed within a false floor (double-layer supported floor) close to the façade in the case of storey-high glazing (17) or within the parapet area. The conditioned air is released into the room through air outlets. Each user can control the unit and can therefore regulate the comfort level in the room to his/her own preference.

14

Simple mechanical ventilation
The simplest mechanical ventilation method is to install small electric fans in a window; a protective screen on the outside provides protection from rain and prevents insects from entering the room. However, this solution is only advisable under special circumstances such as for a server room, where it is not necessary to constantly monitor the operation. The image also shows how dirty the screen becomes over time.

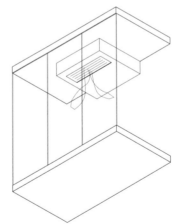

15

Mechanical ventilation by extraction
Exhaust vents are installed in the suspended ceiling. The exhaust air is led through the ducting in the suspended ceiling to the central air-conditioning system.

16

Decentralised ventilation units
Small decentralised ventilation units are installed within the façade grid as necessary. Fresh air enters the building through inlet openings in the façade.

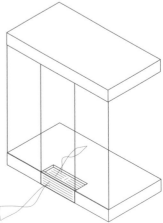

Heating

Ventilating a room causes the room temperature to drop every time fresh air is introduced so to keep the room heated, ventilation causes a constant energy demand. In addition, heat is lost through the enclosing room surfaces which again necessitates heating. The following section describes different options to heat a building with components in close proximity of the façade. (Heating through the room air, which requires a ventilation piping network and central air heating system or an air-conditioning unit is not included here, as they are regarded as being part of general building services).

Heating elements

The most simple and common method of heating a space is using heating elements. Heating elements can be divided into systems based on radiation (18) and those using convection (19, 20).

17

Decentralised air-conditioning unit
Decentralised air-conditioning units were installed in the false floor of the Post Tower in Bonn, Helmut Jahn, 2003. The image shows a unit during installation; the air supply ducts and the water pipes providing the device with thermal heat are clearly visible.

18

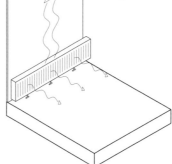

Radiator
A radiator releases part of the generated heat into the room by radiation with a smaller portion being delivered by convection.

19

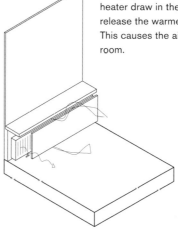

Convection heating
Convection heaters use the principle of warm air rising upwards. The heated fins of a convection heater draw in the cold air from beneath and then release the warmed air upwards into the room. This causes the air to 'roll' to the far end of the room.

20

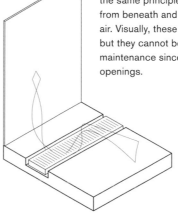

Under-floor convection heating
Under-floor convection heaters are based on the same principle. They, too, draw the cold air from beneath and release it upwards as warmed air. Visually, these systems are less obtrusive but they cannot be adjusted and require more maintenance since dirt can fall directly into the openings.

Façade heating

Another heating method is façade heating whereby warm water runs through the façade sections (21) which improves the comfort level in the vicinity of the façade. In warm weather this system can be used for cooling by using cold instead of warm water. The façade sections filled with warm water radiate the heat into the room; the human body senses this as more comfortable than convection currents which can feel draughty. However, façade heating cannot be used to heat an entire space. More often, it is employed to prevent condensation on the glazing of large façades in foyers or entrance halls. When planning façade heating all connecting points must be carefully examined because the façade elements undergo large linear thermal expansions on the exterior side and at the heat conveying elements.

For the sake of completeness, we should also mention floor heating, panel heating and activated building components. With floor heating, panel heating or activated building components water pipes run through the floor, the wall or loadbearing concrete ceiling units. In the same manner, cooling can be implemented in walls or ceilings.

Cooling

Air-conditioning systems of various types and models are the most commonly used devices for cooling. However, there are also structural measures that can be employed to lower temperatures during warm weather periods to an acceptable level, especially for office and administrative buildings with large expanse of glazing. Such measures can generally be divided into different types of functionality. We can either cool the warm air already existing in a room or use sun protection in or on the façade to prevent incident sunlight from heating up the room air in the first place.

Nighttime cooling

Several principles of cooling can be employed. When using air as the cooling medium the inherent storage capacity of high-mass components can be utilised. When working with a frame construction the floor-ceiling units are the only components that can be used for this purpose but they do provide large usable areas. Concrete ceiling slabs can store thermal energy up to 50-70 mm deep. During the hot summer season, several motorised windows are opened at night when the air has cooled off (23). The cool air then flows along the underside of the rough concrete ceiling slabs (nighttime air cooling). The warm air stored throughout the day is extracted and the cooled ceiling does not heat up as rapidly; the room therefore remains cool for a longer period of time (22).

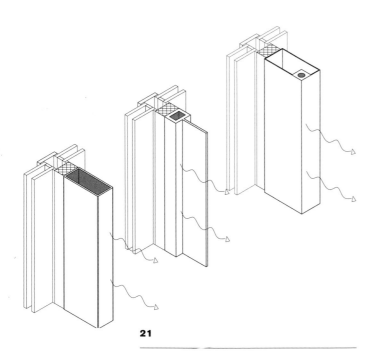

21

Façade heating
Façade heating uses warm water running through the façade sections. The entire façade radiates the heat into the room and improves the comfort level in its vicinity.

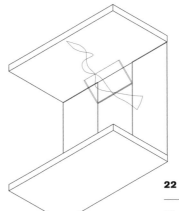

22

Nighttime cooling
The temperature of the ceiling can be lowered by letting cool night air enter through open ventilation flaps so that it flows along the underside of the fair-faced concrete ceiling. During the day the cooled ceiling extracts heat from the room air causing a natural cooling effect.

Activated building components (Chilled ceilings)

Another suitable medium for cooling is water. Water-bearing pipes run through concrete ceiling slabs and are filled with cold water during the hot summer months. The cold water along with the activated mass of the building component extracts heat from the room. An added benefit of this system is that it can also be used for heating by feeding warm instead of cold water into the same pipes (24).

Cooling ceilings

Cooling ceilings are based on the same operating principle but are mounted underneath the structural ceiling or installed as part of a suspended ceiling (25); allowing for future de-installation or upgrades. An alternative solution is the use of cooling wings suspended from the ceiling. By passing cold water through these wings warm air is extracted from the room (26). They can be combined with light fixtures and acoustic elements. However, cooling wings have to be arranged depending on the location of the work spaces, which reduces the flexibility within the room.

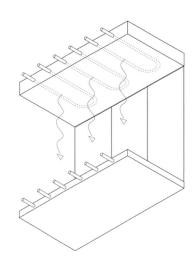

24

Activated building components (Chilled ceiling)
Water-bearing pipes run through concrete ceiling slabs. They can be used with either cold or warm water for cooling or heating, respectively.

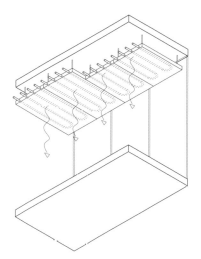

25

Cooling ceilings
The cooling ceiling is suspended from the concrete ceiling and comprises water-bearing elements that radiate cool air into the room. The benefit of this system is that it is easy to install and retrofit.

23

Façade flaps
Flaps are located in the upper and lower areas of the façade elements. The protruding ceiling overhang provides protection from rainwater and allows for unattended ventilation at night.

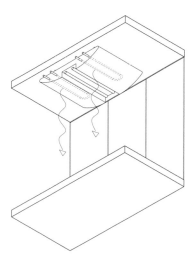

26

Cooling wings
Cooling wings often provide a good cooling solution since they can be combined with light fixtures and acoustic elements.

Sun and glare protection

An energy-efficient adaptive façade can adequately react to the different thermal requirements caused by the changing of seasons. In terms of solar energy input, the requirements on the façade are diametrically opposed during summer and winter. In winter solar energy gain is desired, necessitating the highest degree of total thermal energy penetration (solar permeability of the façade). During summer, however, overheating due to incident sunlight has to be avoided. Besides other passive measures this primarily requires a correspondingly low degree of thermal energy penetrating through the façade. These opposing demands can only be fulfilled by a façade system that can change its permeability of solar energy. This flexibility can be achieved by installing sun protection.

Potential shading of the building caused by structures in its vicinity needs to be examined during the initial design stage. In an urban environment, it might not be necessary to provide sun protection because neighbouring buildings (27) shield off the sun. The following section provides an explanation of the operating principles and place of installation of various sun protection systems.

Interior sun and glare protection

The effectiveness of interior sun protection cannot compare with exterior sun protection. Once the thermal energy has penetrated the glass and entered the room, interior sun protection (28) can block only a small portion of the thermal radiation. However, interior glare protection by means of a screen is effective, specifically for computer work places (29). The screens are made of foils or textiles available with different light transmission values, i.e. the amount of light that can penetrate.

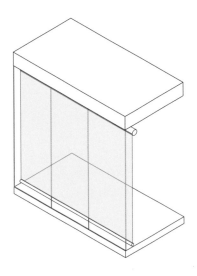

28

Interior sun protection
Interior sun protection is not as effective as exterior sun protection. It is therefore primarily used as glare protection for computer work places.

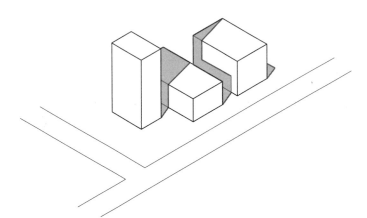

27

Shading
In an urban environment it might not be necessary to provide sun protection on all sides of the building because neighbouring buildings shade parts of the façade.

29

Glare protection
The façade comprises of glare protection blinds on the interior in addition to exterior textile blinds for protection from the sun. The parapet area comprises operable windows and ventilation flaps.

Exterior sun protection

Exterior sun protection systems can be divided into fixed and movable systems. These systems provide the best sun protection because the solar radiation is intercepted in front of the glazing before it penetrates into the building.

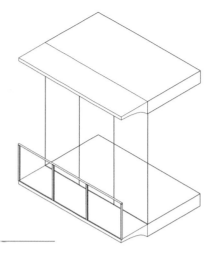

30

Brises-soleil
These fixed sun protection elements protrude from the ceiling-floor units and can serve as service platforms if they have adequate width and carrying capacity.

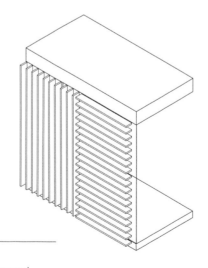

31

Fixed louvres
Fixed louvres can be arranged vertically as well as horizontally. Depending on the configuration the louvres can be adjusted by angling to improve shading. The method of cleaning the glass surfaces behind the louvres needs to be considered during planning.

Fixed sun protection

Fixed sun protection provides a good opportunity for shading. Horizontal elements mounted at ceiling level that protrude far out of the façade are known as brise-soleil (30). Another solution are fixed or pivot-mounted louvres mounted onto the façade. However, they do not achieve the same protection values as those that can be adjusted by angling. And the method of cleaning the glass panes behind the louvres (31) needs to be considered at an early stage.

Fixed sun protection systems can serve alternate functions such as service platforms or secondary emergency exits if they are far enough apart from the glazing.

Using plants is another method of providing fixed shading (32). Deciduous plants are the best choice as they lose their leaves in the winter which increases the possibility of thermal energy penetrating the building during the heating period. However, we need to consider that plants have to be trimmed regularly to avoid obstructing the view and that an irrigation system should be planned for.

32

Façade plantings
Plants arranged in front of the glass panels can be used as fixed sun protection. When using plants as sun protection, regular trimming and irrigation have to be taken into account.

Movable sun protection

One example of movable sun protection are textile systems; either as roller blinds (33) that can be rolled up or down in front of the glass panes or canopy-like systems attached to rails on the façade. Both allow uninhibited outside views.

Another and often used variant are Venetian blinds that comprise of adjustable louvres to regulate the incident sunlight (34). These blinds can be divided into separate parts so that, for example, the upper slats are in a shallower position than the lower ones, allowing sunlight to penetrate deep into the room. Today motorised systems are the norm; some even use sensors to regulate the louvres automatically depending on the position of the sun.

Movable sun protection systems can be damaged by strong winds and therefore need to be retracted during adverse weather conditions. Hence arranging the sun protection inside the façade spacing is an efficient solution, particularly for high-rise buildings since they are usually subjected to high wind loads.

Over the past few years sliding sun protection shades have also been used for apartment buildings and low-rise office buildings (35). The suspended panels can be moved automatically or manually. The frames can be filled in with aluminium or wooden louvres or metal mesh. Due to the horizontal movement of these panels a so-called park position has to be planned for in which the sliding shades can be parked when open.

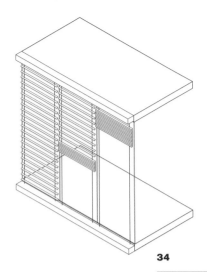

34

Venetian blinds
Venetian blinds are used extensively as sun protection. The market offers numerous systems of varying colours, types of construction and dimensions. Some Venetian blinds include a separate upper area with shallower louvre positions, allowing the sunlight to be directed deep into the room.

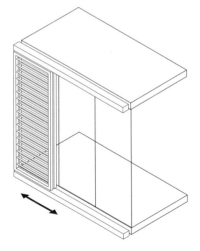

35

Horizontal sliding shades
Horizontal sliding shades can be used in low-rise buildings. They can be motorised or controlled by hand. A wide range of infill materials such as metal mesh, grids, wooden slats or textiles offer many design options.

33

Roller blinds
Textile roller blinds can be moved up and down on cable guides. The type of fabric used determines the degree of visual contact to the outside.

Customised solutions

The industry constantly develops new sun protection products. Some remain as special solutions whereas others mature to standard systems. The following section presents some of the products available today as principle solutions.

Today, sun protection elements are often placed in the spacing between the glass panes of double glazing (36, 37). However, inserting elements in the spacing between glass panes has advantages as well as disadvantages. Inserting metal grids or wooden louvres, for example, saves time and eliminates the need for cleaning. But if a glass pane breaks the sun protection element has to be replaced as well. Motor-controlled Venetian blinds inside double glazing are critical. If a motor fails or the blind jams, the entire double glazing has to be replaced. In general, all benefits and downsides of a specific system need to be considered before making a choice. For example, this solution could make sense if it simplifies the cleaning of the façade. On the other hand separate systems allow for more independence during future operation.

One very simple method of sun protection is to imprint the glass surface with silk-screened patterns of ceramic-based paints that consist primarily of pigmented glass particles called frit (38). This only affects the glass pane itself. Graphic elements of any pattern or grid can be applied to the glass to reduce the incident sunlight. Since this method offers a wide range of possible variations the sun protection can be adapted to the requirements of the specific usage.

37

Central Library, Seattle
In Seattle's new library an expanded metal mesh layer was inserted in the glass airspace as sun protection. For this building with its inclined façade and roof panes, the in-built sun protection is a better solution than external systems.

38

Fritted glass
Using fritted glass is a simple method of sun protection. It provides great design flexibility for the façade. The image shows a façade with several densely fritted glass panes.

36

Central Library, Seattle, OMA and LMN Architects, 2004
This detail shows the rhombic steel beam loadbearing structure of the façade. An expanded metal mesh layer was inserted in the glass spacing to filter the sunlight.

Light-directing systems

Quite often, natural lighting is insufficient for very deep rooms, especially if work places are located on the far side of the façade. In these cases systems that direct the light into those areas can be used. Such systems can also be employed to provide glare-free lighting in the case of direct incidence of sunlight.

Daylight-directing systems can reduce the energy consumption caused by artificial lighting and improve the visual comfort. However, light-directing systems are still very expensive and, from a design standpoint, look very different than other parts of the glazing because they do not offer uninhibited visual contact to the outside. Therefore sun protection systems such as Venetian blinds with light-directing functionality in the upper section are often used instead. In this case, the upper louvres are adjusted at a different angle or they are shaped differently.

Light-directing systems work in different ways. There are horizontal elements that direct the light by reflection (39), and there are those that are vertically inserted into the sun protection system or the glass layer (40). These elements do not reflect the light but re-direct it at a different angle. Many solutions are available, all based on this principle. To name only a few: holographic foils, fine prismatic surfaces and reflective louvres arranged in specific geometries.

The ceiling finish is particularly important for directing or re-directing light because it can support or inhibit light distribution. Simple solutions include painting the ceiling white or mounting light-directing elements.

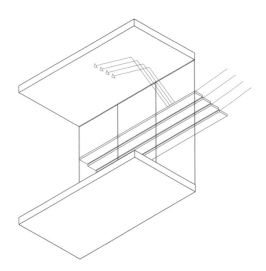

39

Horizontal light-directing systems
Horizontal light-directing elements are mounted on the exterior side of the façade in the form of small consoles. The incident light is reflected and directed across the ceiling toward the far end of the room.

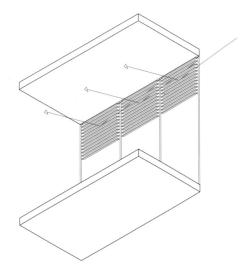

40

Vertical light-directing systems
Vertical light-directing systems installed between the glass panes refract and then spread the light into the room by holographic foils or prismatic glazing. These systems should be installed above the viewable area up to the ceiling to achieve optimum light intensity without restricting the visual contact to the outside.

6 | Adaptive Façades

Buildings able to adapt to changing climatic conditions are called intelligent buildings. Since the term intelligent can be misleading when used in the context of buildings or façades, we will use the term adaptive façade instead. Adaptation generally means that buildings and façades adapt to current weather conditions.

Instead of shutting the environment out, it makes more sense to make use of it since this will have a positive impact on the comfort level of the occupants as well as on the energy consumption. In Central Europe this technology has evolved from traditional methods of construction because the moderate climate allowed for window ventilation and exterior sun protection. Another factor would be shallow building depths, providing each occupant with access to the façade and therefore daylight and natural ventilation.

The downsides of this construction method are high room temperatures during heat waves, susceptibility to wind damage to exterior sun protection (this is especially true for high-rise buildings or in windy regions), draught caused by natural ventilation in winter, and reduced daylight with limited transparency when the sun protection is in use.

Well thought-out concepts for an adaptive façade promise to minimise at least some of these disadvantages so that a high comfort level can usually be achieved even without air-conditioning. In so-called hybrid buildings the comfort level can be further increased by integrating a supporting air-conditioning system for extreme climatic conditions.

The following section provides a short overview of adaptive façade systems. In order to explain the active principle of façades we will begin with clarifying basic aspects of the building physics.

Sun

The sun is the most important supplier of energy and the generator for all living things on earth (1). Compared with the energy demand of the earth the sun's energy reserves are unlimited. The amount of solar energy reaching the earth as solar radiation is only a fraction of the energy travelling to our planet. A portion of the most dangerous solar radiation is blocked by the earth's atmosphere, which forms a safe, filtering layer around the earth. Another part of the solar radiation reaching the earth is absorbed by the water mass of the oceans. The majority of solar energy, however, is absorbed by plants; they convert the carbon dioxide in the air into oxygen by photosynthesis and create the basis for all human and animal life. As primary energy transferring medium, wood, as well as oil and coal produced by the transformation of biomass at the deeper layers within the earth's crust, are used for heating.

Solar energy has only been directly used for heating for a few hundred years, and the technology of solar energy generation is a lot newer than that – it is only a few decades old.

Related to buildings the sun can be utilised as a generator for cleverly devised climatic concepts and for facilitating natural air circulation.

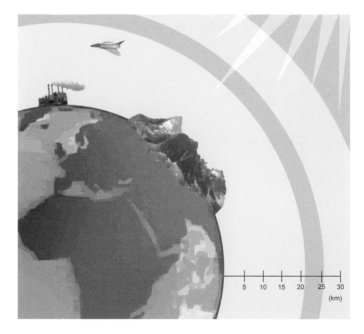

1

Earth's atmosphere
The earth is enclosed by several layers of air that filter off hazardous spectrums of the solar radiation, thus blocking them from the earth. Compared with the energy demand of the earth, the sun's energy reserves are unlimited. Only a small portion has been used so far.

Light

Light is the term for the range of electromagnetic solar radiation that can be perceived with the human eye. This range includes wavelengths of between 380 and 780 nanometres (2). Below this visible wave range is the short wave ultraviolet radiation – the long-wave infrared radiation lies above. If the energy of long-wave radiation is increased, it can even be used as the cutting force of a laser beam. The human eye perceives the stimuli and intensity of incident light as shades and colours.

Heat

Heat or heat quantity is a physical value and describes the transmission of thermal energy across system boundaries. Heat is tied to this transmission process and is therefore a process factor, as opposed to a state variable. Thereby thermal energy is always transmitted from the system with the higher temperature in the direction of the system with the lower temperature (3).

When heat radiation like sun-rays hits an object it can either be partially transmitted (transmission), partially reflected (reflection) or partially absorbed (absorption). If an object absorbs the heat radiation, it heats up. This heat can be passed on according to the principles of heat transmission.

Within a substance this is called thermal conduction. Within a fluid medium, including air, heat spreads in the form of convection, whereby warm air becomes lighter because its density decreases, and rises, resulting in an air current. Transmission of heat from object to object is called radiant heat or radiation. The thermal flow is always directed from the higher temperature level to the lower level.

The phenomenon of thermal radiation is very noticeable when we sense the heat of the sun on our skin or feel uncomfortable standing next to a cold wall because it draws the warmth from our bodies.

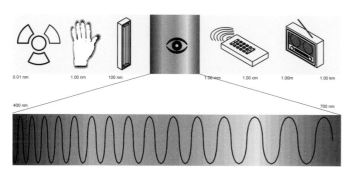

2

Visible light spectrum
The human eye can only perceive a small portion of solar energy as visible light. The wave spectrum ranges from short-wave ultraviolet radiation to long-wave infrared radiation.

3

Heat transmission
Heat can be transmitted in different ways; the energy is transported or dissipated depending on the medium. The different transmission mechanisms can be categorised as radiation and thermal conduction.

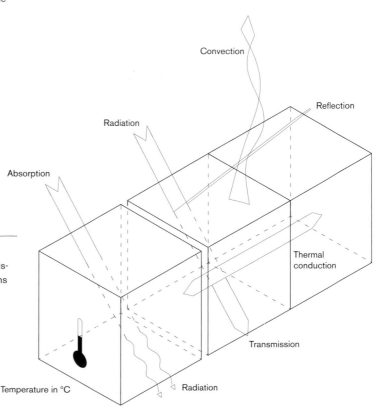

Greenhouse effect

The principles mentioned above help to describe the greenhouse effect, the reason why glass is used in architecture to utilise solar energy.

If short-wave radiation hits a glass pane it can penetrate the glass. So glass transmits the solar radiation; the radiation subsequently hits objects such as the floor or walls of a room and is absorbed. During this process the radiation is transformed into long-wave heat radiation. This heat radiation is then transmitted into the room (4). Unlike short-wave solar radiation, this long-wave heat radiation cannot penetrate glass, causing the room to heat up. Numerous concepts to heat and naturally ventilate buildings can be derived from this one-way transmission property inherent to glass.

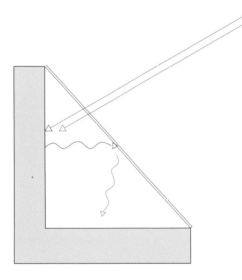

4

Greenhouse effect
Since short-wave radiation penetrates through glass, objects behind the glass layer can heat up by absorbing the radiation. Short-wave radiation is then transformed into long-wave radiation for which glass proves impermeable. The room heats up accumulatively.

History of adaptive façades

Within the scope of the technologies of their time, traditional farmhouses had already made optimum use of energy-saving potentials. The heat generated by the livestock was used for heating the building, and straw and hay were not merely bedding and feed but provided insulation. Energy consumption caused by burning firewood was kept to a minimum. Windows had folding shutters that created a thermal buffer between the glass and the shutter at night, very much like a double-skin façade today (5).

In alpine regions, this method of construction still exists today. Double-skin structures make up one of the most widely employed functional principles used to protect against exterior environmental influences through the façade envelope. Prior to the development of insulated glass a second window was installed to utilise the area between the two windows as a thermal buffer. The combination of two single glass panes in this box window (6) generates higher insulation values and can be adapted to the prevailing weather conditions. During winter both windows remain closed, whereas during summer the exterior windows can be opened to provide ventilation.

5

Historic half-timbered house
The windows employ timber folding shutters to adapt to changing weather conditions. The layer of air between the glass and the folding shutters serves as thermal insulation.

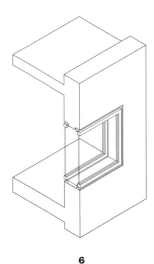

6

**Functional principle
of a box window**
Schematic drawing of a box window
with two individually operable windows.

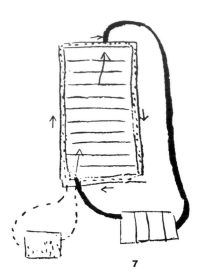

7

Mur neutralisant
Le Corbusier already developed the
idea of a climatically active façade
that actively shields the building from
exterior influences, in order to achieve
comfort for every climate.

In modern times glass has been used more and more frequently; however, this has increased the issue of excessive cool-down in winter and overheating in summer. As early as 1929 Le Corbusier formulated a concept for a building envelope with positive impact on the indoor climate (7) in *Precisions: On the Present State of Architecture and City Planning*. He talked about the 'mur neutralisant': 'We have seen that these neutralizing walls are in glass, in stone, or in both. They are made up of two membranes with a space of a few centimetres between them. [...] A circuit in that narrow interval between the membranes, hot air is pushed if in Moscow, cold air if at Dakar. Result: One has regulated in such a way that the inside face, the inside membrane, stays at a temperature of 18 degrees. There you are! [...] The house is sealed fast! No dust can enter it. Neither flies nor mosquitos. No noise!' (*Precisions*, p. 66)

Le Corbusier's thoughts were never conveyed into a satisfactory result. His ideas were far ahead of his time. Today, his 'mur neutralisant' can be seen as the predecessor of the exhaust-air façade; this type of façade allows regulating the environment of the usable spaces individually, independent of the exterior environment by employing a combination of a double-skin structure and an air-conditioning unit.

Whereas Le Corbusier aimed to moderate the room adjacent to the façade with an artificial environment in the building envelope (independent of the exterior conditions), modern environmental concepts use the gap between façade layers to create a buffer. Thus the façade space creates an intermediate environment between the interior and the exterior.

Another concept defines the façade as a regulating layer similar to a filter between the inside and the outside which allows an exchange of environmental conditions through the façade, depending on the requirements. Whereas Le Corbusier's mur neutralisant was based on the idea of actively neutralising exterior influences on the interior space, these façades utilise exterior environmental conditions and make them available for optimum use by the building. This type of façade, known as a collector façade, employs environmental energy by mostly passive means. This façade also incorporates a buffer zone within the façade layers; but unlike the concepts described before it interacts with the exterior climate within the outer shell.

In the forties, Buckminster Fuller had already developed concepts that comprised of a dome-shaped structure as a secondary envelope to generate an independent microclimate with passive means alone. The effect of wind and sun on the envelope were to be the only methods to generate cooling, ventilation and heating. Buckminster Fuller, Norman Foster and Frei Otto also considered the use of a large environmental envelope to create a microclimate, much like a cheese cover.

None of these visionary ideas were ever realised but drove the development of solar architecture in the U.S. during the sixties and seventies. These were mostly ecological detached houses, self-constructed with solar façades and solar collectors.

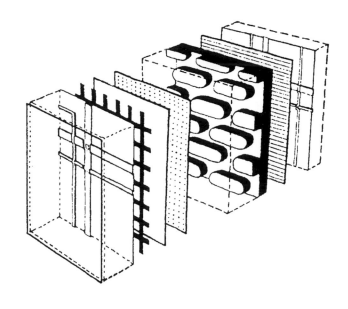

8

Polyvalent wall
In 1981 Mike Davies formulated the idea of a polyvalent wall in which all the functions of the façade were to be fulfilled by one element with several layers. The façade was also meant to generate the necessary energy itself.

In 1981, working for Richard Rogers and Partner, Mike Davies already formulated the idea of a polyvalent wall (8) in his article titled 'A wall for all seasons'. Here, several functional layers within a glass element were to provide sun and heat protection, and to regulate the functions automatically according to current conditions. The wall itself was to generate the necessary energy. The term 'intelligent façade' derives from the concept of the polyvalent wall. Although the matter of technical realisation has not yet been resolved, the polyvalent wall is the vision as well as a driving force for new façade technologies, and many scientists have been engaged in this topic over the last two decades.

The oil crisis in 1972 and the resulting awareness that resources are limited led to considerations of using the energy created by solar radiation incident on façade surfaces. The ecologically conscientious building movement in the eighties was a consequence of these developments. The following section describes the main types of façades that actively create energy, known as collector façades.

Collector façade
Trombe wall

The Trombe wall (9) is the most simple collector wall and uses the greenhouse effect. Short-wave sunlight penetrates the glass panes on a south-facing wall and hits on a dark absorbent layer – the so-called Trombe wall. It is absorbed and transformed into long-wave heat radiation. The heat in the gap between the façade layers is transmitted through the wall into the room behind it. Depending on the structure of the wall and its storage capacity, the heat gained can be discharged quickly or over a long period of time, well into the evening hours. If there are openings at the top and the bottom of the wall, then the thermal difference within the gap causes the room air to circulate (10). Cold air is drawn into the gap at the bottom, heated up and then exits into the space behind. This principle is called air heating. If additional openings are installed in the exterior glass layer, the air circulation within the gap feeds warmed fresh air into the room. The same principle applies to leading exhaust air out of the exterior façade.

One example of a simple Trombe wall is the residential house of Steve Baer in New Mexiko, U.S.A., built in 1973 (11, 12). A wall of (water-filled) oil barrels stores the heat of the sun during daytime. The wall is insulated from the inside space by a cover. At night, the exterior covers are closed and the interior covers open, so that the heat stored during the day can be discharged into the room.

9

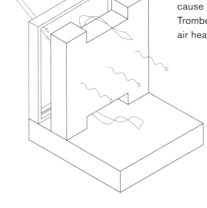

Trombe wall
The Trombe wall utilises the greenhouse effect. Solar energy penetrating through the glass is absorbed by the thermal mass of a dark wall and later discharged into the interior space.

10

Ventilated Trombe wall
Vent openings in the Trombe wall cause additional convection. The Trombe wall can then function as an air heating system.

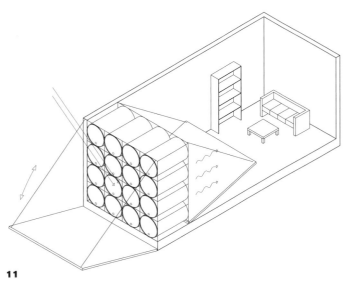

11

Detached house, Corrales, New Mexico, Steve Baer, 1973
Functional principle of Steve Baer's Trombe wall made from oil barrels. During the day, solar energy is stored in the water-filled oil barrels. At night the exterior covers are closed and the interior covers are opened, so that the heat can be discharged into the room.

12

Detached house, Corrales
The image shows the oil barrels in the open façade as well as the tackles used to operate the covers from the inside.

Transparent heat insulation

Transparent heat insulation (THI) should be mentioned in this context. THI elements (14) can be used instead of the single glass layer. THI elements are installed in front of the absorbing wall (Trombe wall), solar radiation penetrates the THI elements and heats the absorbing wall while the THI elements minimise heat loss toward the exterior. However, in order to prevent over-heating in summer, sun protection has to be installed in front of the THI. Bevelled glass is being developed that would reflect the sun beams when the sun's position is high (summer), so that the solar energy is only absorbed during the cold season when the sun is lower in the sky. THI elements are useful in combination with a collector wall to utilise the principle underlying the Trombe wall and to allow for adaptation to the seasons. THI elements can also be used alone without a collector wall to light the room with diffused light and to improve heat insulation at the same time.

Transparent heat insulation can be based on different operating principles whereby the geometric arrangement of the THI layer varies. All THI elements increase heat insulation and let diffuse light enter the room, depending on the method of construction. In order to protect the materials used, they are all installed between two layers of glass. We can differentiate between four applications (15).

1 Light is guided through standing air layers parallel to the glass layer. The light is reflected by the individual separating layers, usually consisting of polycarbonate or Plexiglas and falls in the inside space.

2 Light is guided perpendicular to the glass layer through honey-comb shaped or tubular plastic elements. One advantage of the vertical arrangement is the formation of small air volumes which increase the insulation value.

3 Closed-cell fills or chamber structures are applied as translucent materials such as acrylic foam fills or saline hydrates. Convection of the enclosed air is largely prevented. Fills of light fibre glass mats are also possible.

4 Almost homogenous structures can be created by using microscopically fine materials such as aerogel. Aerogel consists of 2-5% silicate and 95-98% air; this slightly porous material is also called 'solid smoke' since it resembles solidified smoke (13).

13

Aerogel
NASA developed aerogel as early as around 1950; it is a very good thermal insulator. The high percentage of air (ca. 95-98 %) makes it extremely light. Due to its appearance it is also referred to as 'solid smoke'.

14

Typical THI elements
Shown here are parallel and perpendicular structures, fibre glass webbing and aerogel as fine structures.

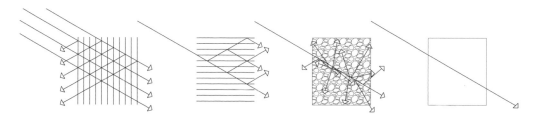

15

Structure of THI elements
According to their functionality, THI elements can be grouped into four categories:
1 Parallel to the glass layer.
2 Perpendicular to the glass layer.
3 Chamber structures.
4 Almost homogenous structures.

Exhaust-air façade

Exhaust-air façades consist of a multi-layered façade structure; however, they differ from double façades in that they require mechanical ventilation to achieve air movement (17). The interior layer is often made of single glazing or textile roller blinds. The enclosure to the exterior consists of double glazing. The exhaust air is drawn through the space between the façade layers via vents. The air can flow in two directions; from the bottom upwards or from the top downwards. The exhaust air is then transferred through air ducts and typically exits through the roof. The exhaust system can include a heat recovery option. The thermal energy generated by the sun protection warming up the gap between the glass panes is extracted through the ventilation system. The continuous circulation of warm air in the façade gap increases the comfort level in the vicinity of the façade. This comfort gain allows for more workstations close to the façade. An exhaust-air façade usually excludes natural ventilation and the building requires year-round mechanical ventilation. Due to the enclosed outer skin, this type of façade achieves good sound insulation. A simple variant of an exhaust-air façade is one without the interior glass layer, using only the gap between the double glazing and the sun protection blind as an air conducting space. Buildings known to employ exhaust-air façades are the New Parliament Building in London by Michael Hopkins and Partner (18) and the Lloyd's Building (16), also in London, planned by Richard Rogers. The air ducts in the façade of the New Parliament are clearly visible – they reach beyond the roof and gather in large chimneys. On the other hand, in the Prisma Building in Frankfurt by Auer + Weber + Partner, 2001, the exhaust-air façade functions as a solar chimney: the exhaust air is extracted from and through the entire façade. It therefore constitutes a variant of the exhaust-air façade with natural ventilation. (However, from a structural point of view this façade is a double façade).

16

Lloyd's Building, London, Richard Rogers, 1986
In the Lloyd's Building by Richard Rogers, the air ducts of the façade are recognisable by the ventilation ducts attached on the outside.

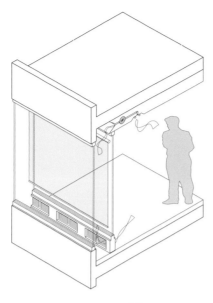

17

Exhaust-air façade
With an exhaust-air façade the exhaust air is extracted through the space between the façade layers by means of a central air-conditioning unit. Therefore the temperature of this space drops only marginally compared to the room temperature, resulting in increased comfort in the area close to the façade.

18

New Parliament Building, London, Michael Hopkins, 2000
The air ducts in the façade of the New Parliament can be traced beyond the roof to large exhaust chimneys.

Double façade

From a structural point of view double façades consist of three functional layers. Typically the exterior façade layer is made of single glazing. It is separated from the interior glazing, which in turn forms the room enclosure and usually consists of double glazing. Depending on the specific design, the distance between interior and exterior façade layer can vary. In order to utilise the effect of a thermal buffer in the space between the two façades, ventilation openings are installed in either one of the exterior and interior façade or in both. The air in the gap between the façades heats up due to solar radiation and hence serves as a buffer toward the interior space. Due to the thermal difference, the warm air can be used as a generator of natural ventilation of the interior room or the space in between the façades.

Double façades are most suitable for protecting the interior space from noise sources such as highly frequented streets. However, when designing a double façade as sound insulation for a specific project, we need to examine the different methods of construction.

In case of strong winds the sun protection elements installed on single skin façades of high-rise buildings have to be retracted to prevent damage. This, of course, entails that the adjacent rooms will not be protected from overheating. Therefore double façades are especially suitable for high-rise buildings since sun protection elements can be safely installed inside the gap between the two façades so that they are not exposed to the wind. Depending on the method used to conduct air in the space between the two façades, double-skin façades can be grouped into four main categories:

• In a box-window façade the air only circulates within one façade element.

• In a shaft-box façade the air rises in vertical shafts alongside the façade and draws the exhaust air from adjacent façade elements with it.

• In a corridor façade the air circulates within the gap between the façades horizontally across one storey.

• In a second-skin façade the air circulates across the entire façade area within the unrestricted gap between the two façade layers.

Box-window façade

The box-window façade is based on the principle of the box window but consists of storey-high façade elements (19). The interior windows can be opened for ventilation into the gap between the two façade layers. The exterior façade comprises openings for supply and exhaust air. Horizontal as well as vertical separation from adjacent elements ensures optimum sound insulation not only from the outside but from neighbouring offices as well. Unpleasant odour and flashover can be prevented rather easily if the compartmentalisation is designed correctly. Thermal shorts, meaning exhaust air from a lower element flowing into an element above, can be avoided by offsetting the supply and exhaust openings from storey to storey.

One example of a box-window façade is the Daimler-Chrysler Building by Hans Kollhoff (20) at the Potsdamer Platz in Berlin. However, instead of using storey-high elements, box windows were inserted in the fenestrated façade made of prefabricated clinker brick components (21). The exterior glazing can be opened for cleaning. Ventilation is provided for by gaps at the top and the bottom, whereby the upper gap can be closed by vertically shifting the position of the exterior window. This could better utilise the air collection effect in the façade gap during heating periods.

20

Daimler-Chrysler Building, Potsdamer Platz, Berlin, Hans Kollhoff, 1999
The high-rise Daimler-Chrysler building at the Potsdamer Platz in Berlin is equipped with box windows.

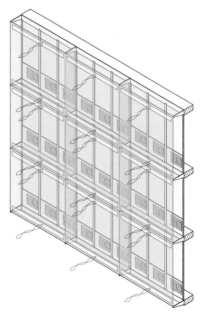

19

Box-window façade
The box-window façade is derived from the box window principle. Horizontal as well as vertical separation makes the box-window façade especially suitable for sound insulation, not only from the outside but from neighbouring offices as well.

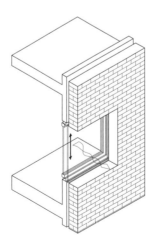

21

Daimler-Chrysler Building, Berlin
The exterior windows of a box-window façade are used to adapt to the climate. In order to do so, they can be moved vertically by opening or closing a ventilation gap in the upper area. This mechanism offers improved thermal buffering during winter and better ventilation during summer.

Shaft-box façade

Shaft-box façades are derived from the same construction principle of the box-window façade (22). Box windows and shaft elements alternate, and the shaft elements extend across several storeys. The vertical shafts are connected to the box windows via overflow openings storey by storey. Due to a stack effect, the warm air flows from the façade gap through openings at the head of the element through the shaft to the outside. The exhaust air can be extracted from the façade gap mechanically. However, the required fan performance would be very high, which usually renders this method uneconomical. During winter, low ventilation increases the buffer effect but this can result in condensation forming on the interior side of the outer glass pane when the interior façade is open.

Due to the uncluttered flow diagram of the shafts, fewer openings are needed in the exterior façade, resulting in a lower demand for sound insulation.

Since the stack effect increases with increasing height, the total height should be limited.

This type of façade is suitable for lower rise buildings. The Photonics Centre in Berlin-Adlershof (23), planned by Sauerbruch Hutton Architects, is an example for this façade type. The storey-high shafts are formed by the loadbearing structure. The exhaust air travels upwards through the shafts and exits at the top of the building through ventilation louvres. A shaft-box façade was used for the ARAG Tower in Düsseldorf (24, 25), designed by RKW Architektur + Städtebau in cooperation with Foster and Partners. This façade comprises four stacks of seven storeys each, therefore reducing the total length of the shaft to a quarter in order to keep the stack effect at a low level.

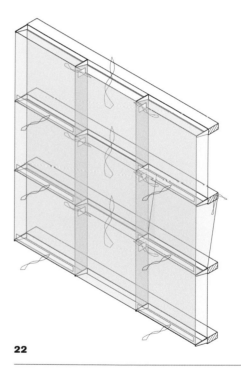

22

Shaft-box façade
Shaft-box façades comprise box window elements and exhaust shafts arranged in alternating sequence within the façade layout. The stack effect inside the shafts causes the exhaust air to be drawn out of the box windows naturally and then exhausted through the roof.

23

Photonics Centre, Berlin-Adlershof, Sauerbruch Hutton Architects, 1998
The ventilation shafts are vented through ventilation louvres at the head of the façade. The image clearly shows the openings in the concrete columns, through which the air circulates, as well as the interior sun protection and the interior façade.

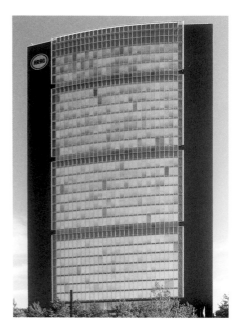

24

___

**ARAG Tower, Düsseldorf, RKW Architektur
+ Städtebau with Foster and Partners, 2000**
This shaft-box façade comprises four stacks of
seven storeys each. Dividing the exhaust system
into four sections limits the air flow within the
exhaust shafts.

25

___

ARAG Tower, Düsseldorf
The continuous vertical exhaust shafts can be
identified by the light reflections. Two box window elements followed by one shaft are arranged
in sequence.

Corridor façade

Corridor façades manage the airflow storey by storey (26). In
some cases vertical dividers are added for fire or sound protection because sound propagation to neighbouring rooms can occur through the gap between the interior and the exterior facade
layer.

Air inlets are located near the floor and the ceiling (29). They
are arranged at an offset to avoid thermal shorts by exhaust air
mixing with fresh air. Separating the individual storeys from each
other effectively prevents overheating that can occur at upper-
level storeys when the air is led across several storeys. This corridor is accessible and is typically designed to be wide enough
to be used as a service platform. The space between the façades is ventilated through openings at ceiling level. The airflow
can be regulated by motorised flaps.

The Stadttor Building in Düsseldorf (27, 28) is one example
where the corridor façade comprises a gap of up to 1.40 m.

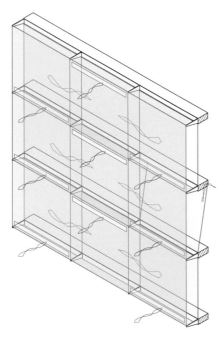

26

___

Corridor façade
In corridor façades the air flows within the space
between the exterior and interior façades across one
storey. Air in- and outlets are arranged at an offset
at ceiling level to avoid thermal shorts by exhaust air
mixing with fresh air.

27

**Stadttor Building, Düsseldorf,
Petzinka Pink und Partner, 1998**
The image clearly shows the ventilation louvres at ceiling level, the deep gap between the two façade layers and the interior sun protection of this corridor façade.

28

Stadttor Building, Düsseldorf
Close-up shot of the intake flap in the space between the façade layers. The image shows the ventilation grid behind the exterior façade through which fresh air enters the space.

29

Stadttor Building, Düsseldorf
The broad accessible space between the façade layers, the ventilation slots at ceiling level and the interior façade are clearly visible.

Second-skin façade

Second-skin or multi-storey façades do not compartmentalise the space between the façade layers. Instead, the exterior façade contains a layer of air that envelops the entire building as a buffer in front of the interior façade (30). The rooms are often ventilated mechanically. The space between the façades can serve as a supply or exhaust air system.

The exterior façade is ventilated through openings at floor and ceiling level. The vents can be closed during winter to make use of the greenhouse effect and to increase the thermal protection. In summer, the façade flaps can be opened to prevent overheating. The limited number of ventilation openings ensures good sound insulation from the outside but, within the façade, entails the risk of sound propagation from room to room. Fire protection is another critical issue because, in case of fire, the smoke spreads quickly throughout the space between the façade layers.

In traditional high-rise constructions, second-skin façades are used as buffer façades with a small gap spacing, and as large environmental envelopes discreet from the enclosed buildings. The façade gap can vary in depth up to its complete disintegration, which then results in a space-forming exterior envelope.

During winter, when ventilation is low, the buffer effect increases; however, the risk of condensation forming on the inside of the exterior façade increases in equal measure. The air quality decreases because fresh and exhaust air mix within the space between the façade layers. To avoid these problems, the double façade can be used as a supply air façade in winter and as an exhaust air façade in summer. This concept was realised in the Prisma Building in Frankfurt by Auer + Weber + Partner (31, 32).

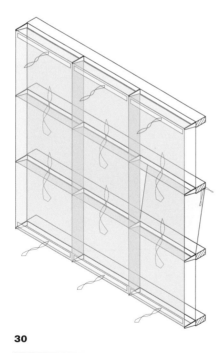

30

Second-skin façade
With a second-skin façade the interior façade is enveloped by an unrestricted glass layer around the entire building. Good sound insulation against exterior noise sources can be obtained because the in- and outlet openings are located only at floor and ceiling level.

31

Prisma Building, Frankfurt, Auer + Weber + Partner, 2001
This double façade serves as an exhaust façade during summer and provides fresh air supply in winter.

Another example worth mentioning is the Double-XX Office Building by Bothe Richter Teherani (34) in Hamburg. Here the building is set far inside the exterior façade, resulting in large atria that form green open spaces in the floor plan.

The Academy Mont Cenis in Herne by Jourda & Perraudin (33) is based on the 'house within a house' principle, which sets several buildings within one large glass envelope. The large volume enclosing the buildings is supposed to provide a moderate environment throughout the year.

33

Academy Mont Cenis, Herne, Jourda & Perraudin, 1999
The academy for continuing education is enclosed by a large glass envelope that functions as a buffer zone for the buildings within. Due to its large volume, the indoor environment can be adjusted uniformly throughout the year.

32

Prisma Building, Frankfurt
The space between the two façade layers is accessible; the façade flaps with operating hinges to open the façade are visible.

34

Double-XX Office Building, Hamburg, Bothe Richter Teherani, 1999
The second-skin façade encloses the entire building. Where the interior façade separates from the exterior façade due to the geometry of the floor plan, spaces are created that range from small gaps between the two façades to large atria.

Alternating façade

Double façades have been built in great numbers; many are documented and have been presented as technical innovations. Today we know their potential but also of the problems related to specific locations or types of use.

As a further development, double façades were combined with single skin façades to create so-called alternating façades (35). By combining these two known construction and functional principles, it is possible to achieve compliance with the given requirements. Sometimes alternating façades are called hybrid façades; the word hybrid (Greek = coming from two directions) describes its technological origins.

Because the double and single façade areas alternate (36), in winter, warm air can be drawn from the façade gap of the double façade to supply adjacent offices with fresh pre-heated air, thus reducing the energy demand for ventilation. In summer, the single façades provide natural ventilation when very warm air from the double façade sections can cause problems. The space in between the layers of a double façade can be ventilated by opening ventilation flaps so that adjacent rooms are not overheated. Alternating façades can be realised as storey-high façades as well as in-line or fenestrated façades.

Integrated façade

Considering the technological development of the façade, which has always been equipped with heating elements in the interior space, and the technological advancements of progressively smaller decentralised air-conditioning and ventilation units, it seems reasonable to integrate these components into a façade module. From the construction point of view it is advantageous to integrate as many components into the façade as possible. The industrial manufacturing process of façade modules makes it possible to integrate more components with high accuracy; they are then mounted on the shell of the building as unit system façades in the proprietary manner. This method reduces the time needed to assemble building services components in the shell of the building.

Today, functions such as heating, cooling, ventilation as well as light-directing, shading, integration of artificial lighting and even energy generation with solar panels can all be realised in integrated façades (38). These functions can be combined on the basis of a modular design principle, giving consultants the option to design the façade according to discreet requirements.

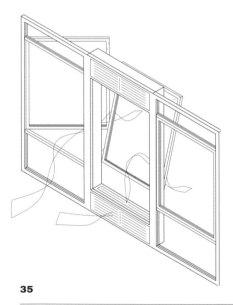

35

Alternating façade
The alternating façade combines a double façade with a single skin façade. In summer, the single skin façade sections provide cooling to counteract possible overheating caused by the double façade. In winter, pre-heated air can be drawn from the space between the layers of the double façade which reduces the energy demand for heating.

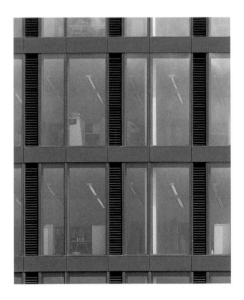

36

Debitel Headquarters, Stuttgart, RKW Architektur + Städtebau, 2002
A weather guard grid is mounted in front of the single skin areas of this alternating façade, allowing the windows to remain open unmonitored for nighttime cooling. The glazed areas in the photo are double façade sections.

On the one hand, the large number of decentralised air-conditioning units raises the maintenance requirements and increases the complexity of environmental control engineering. On the other hand, cost savings are achieved with regards to the central environmental control units, shafts and ducting as well as lower storey heights because horizontal air flow is typically not required. Individually-adjustable room environment and air quality present additional benefits because they increase the comfort level.

The façade of the Capricorn House in the Medienhafen Düsseldorf by Gatermann + Schossig (37) is a good example of this façade principle. The façade integrates decentralised air-conditioning units within enclosed façade sections that provide cooling, heating, ventilation and air-conditioning. The units also comprise heat recovery systems that extract the energy from the warm exhaust air and use it to pre-heat the fresh air supply. Furthermore, daylight-directing louvres are installed in the fan light area to increase the daylight in the room. Light fixtures are installed in the façade elements, providing direct and indirect lighting.

When examining the different types of adaptive façades that include increasingly specialised functions and components, it becomes apparent that façades are becoming more and more complex. Whereas during the initial development stages many physical innovations such as natural ventilation in double façades were realised, latest enhancements show a significant increase in building services-related components.

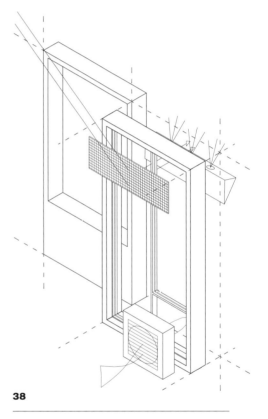

38

Integrated façade
An integrated façade comprises numerous building services elements. The building process can be shortened because additional components can be integrated into the façade elements during the industrial manufacturing process.

37

Capricorn House, Düsseldorf, Gatermann + Schossig, 2006
The integrated façade of the Capricorn House comprises integrated air-conditioning units behind the opaque areas of the façade, allowing for individual adjustment of the room environment.

Rear-ventilated façade

Project | Concept House, RDM Campus
Location | Rotterdam, the Netherlands
Completion | 2012
Client | TU Delft
Architect | TU Delft
Structural engineering | TU Delft
Building services | TU Delft
Façade planning | TU Delft

Conception

Concept Car – Concept House! The Concept House on the RDM Campus in South Rotterdam is a test building, developed by the Chair of Product Design of TU Delft. (The RDM Campus is located on the grounds of the former dockyard Rotterdamsche Droogdok Maatschappij, which gave the campus its name.) The objective was to develop a building system for an urban villa with four storeys, four flats per storey. It was designed to allow for quick assembly in the form of a simple plug & play system. With careful planning and prior coordination, all essential building functions were identified and integrated into individual components, which could then be assembled on site as prefabricated building elements. Another objective was to gain energy neutrality with roof-mounted photovoltaic cells.

1

Concept House
The realised flat, a module for larger residential dwellings, is built on stilts.

2

View of the façade
Façade of the test building with a folding sliding wall.

The project (1) realised on the Rotterdam test area consists of one flat with one living room and kitchen, two bedrooms and a sanitary unit. Since it was designed modularly and for the purpose of easy transportation, the body of the building consists of a timber skeleton construction with insulating infills and a rear-ventilated façade. To emphasise the impression of a module for future residential building in general, the realised project – consisting of only one flat – is set on stilts. For this prototype a maritime container functions as a basement.

Façade

The façade consists of a modular timber skeleton construction, developed as a prefabricated system into which timber windows were inserted. The front side along the living room area was enclosed with a glass folding sliding wall in the form of a timber aluminium system (2). The timber skeleton construction consists of vertical timber members (TGI joists) and cellulose insulation, clad on the inside with an OSB panel. In certain areas, an additional installation layer of gypsum cardboard is added to prevent penetration of the vapour-tight layer when installing electric installations. The outer layer sequence consists of a sheet of wood products, a vapour-permeable wind barrier and a 70 mm thick rear-ventilated cladding system made of aluminium rails and ceramic panels (3). The ceramic cladding overlaps the joints of the building elements. This particular material was chosen to hide the modular character of the building. The cladding is mounted on site following the installation of the prefabricated parts (4).

3

Detail of ceramic cladding
The rear-ventilated ceramic cladding is mounted on a support structure.

4

Corner detail of the façade
Corner solution of the rear-ventilated façade with ceramic tiles mounted on a support structure of horizontal and vertical aluminium sections.

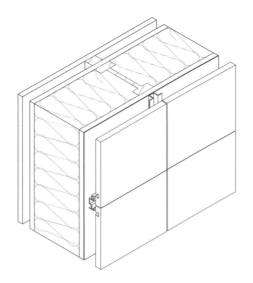

5

Isometric view of the timber skeleton construction
Façade consisting of storey-high timber members, filled in insulation as well as the interior OSB layers and the exterior sheet of wood board with wind barrier. The façade is enclosed with a rear-ventilated ceramic cladding.

An isometric view (5) shows the timber skeleton construction. As described earlier it consists of vertical TGI joists. The upper and lower ends of the system feature timber planks to create a simple connection and to ensure safe load transfer downward. On the inside OSB panels are mounted onto the vertical members. These panels also serve as a vapour-tight layer. An internal installation layer made of gypsum board, commonly used in timber construction today, allows for a 35 mm deep space for electric installations without having to penetrate the vapour barrier. Cellulose was chosen as insulation material, installed with a thickness of 25 cm (6).

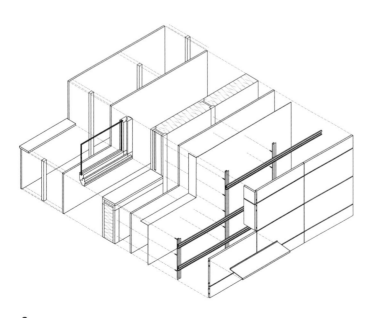

6

Exploded isometric view of the timber skeleton construction
The individual layers of the timber skeleton construction are clearly visible. From inside to outside: gypsum cardboard, counter battens, timber skeleton with TGI joists with insulation, sheet of wood board, wind barrier, non-load-bearing rear-ventilated façade cladding with aluminium support structure and façade panels.

7

Detail of the ceramic cladding
The outer envelope is secured with special brackets onto which the ceramic tiles are mounted with undercut anchors.

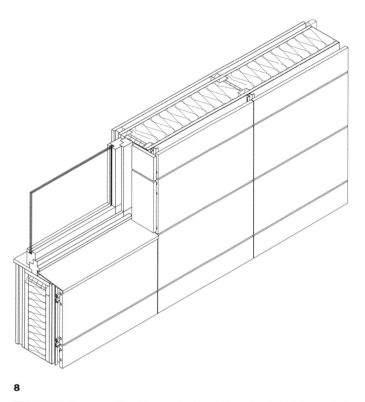

8

Isometric view of the façade detail
Construction detail with timber posts in the centre, the individual layers and the exterior cladding.

The outer layers consist of a sheet of wood products, the vapour-permeable wind barrier and the exterior rear-ventilated façade system. The façade system in turn consists of a vertical supporting structure and horizontal aluminium sections, which hold the individual ceramic tiles. The fixtures behind the façade panels – which are fastened with undercut anchors – are hooked into the horizontal rails. They are thus invisible and allow precise adjustment to accommodate tolerances and ensure an accurate joint pattern (8, 10).

The construction of the ceramic cladding can be seen in the exploded view (9). The exterior rear-ventilated cladding consists of ceramic tiles made from dyed through porcelain stoneware, which does not show any colour variation even if damaged. The tiles are fixed to special brackets with undercut anchors (in conical drill holes). The brackets in turn are mounted on the support structure. The upper brackets are equipped with levelling screws to compensate for tolerances. The lower anchor has sufficient play to avoid introducing tension into the tiles in case of thermal expansion (7).

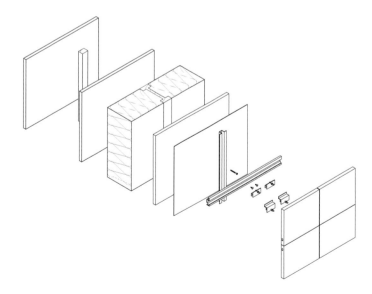

9

Exploded isometric view of the façade detail
Geometric dissolution of the node with the individual constructional layers consisting of the loadbearing support structure, the insulation layer, an inner OSB panel and an exterior sheet of wood board with wind barrier and the rear-ventilated cladding.

10

Window connection
Connection detail of the window with an already mounted window and rear-ventilated cladding.

Solid façade

Project | State Archive Nordrhein-Westfalen
Location | Duisburg, Germany
Completion | 2012
Client | Bau- und Liegenschaftsbetrieb NRW
Architect | Ortner & Ortner Baukunst
Structural engineering | office for structural design (osd)
Building services | Arup GmbH
Façade planning | Gödde Architekt

Conception

The State Archive (Landesarchiv) Nordrhein-Westfalen was designed as a reconstruction of an existing storehouse from the 1930s in the Duisburg harbour. The design revolves around the 75 m high archive tower, which houses the archive across 21 storeys. The tower was integrated into the existing building stock and towers over the old storehouse (1).

From a constructional viewpoint, the storehouse and the new archive tower are divided through separate foundations. The storehouse consists of a concrete shell that carries the ceilings and is clad with a masonry shell. In order to emphasise the monumentality of the building, the steel structure of the roof is also clad with bricks. In those areas of the old storehouse that house part of the archive the old windows were filled in with masonry; however, they are still recognisable in the façade (2). A six storey high flat roof building adjacent to the archive (3) houses the administration and reading rooms. Here, the façade was realised in the form of a thermal insulation composite system with ribbon-shaped windows in order to save cost.

2

Gable State Archive Nordrhein-Westfalen
The materiality of brick is emphasised by the different masonry patterns.

1

State Archive Nordrhein-Westfalen
The storehouse, located in the old harbour, was expanded by a high archive tower.

3

Office annex
The annex contains reading rooms and functional areas in addition to the offices.

Façade
The existing storehouse consists of solid masonry made of historic bricks in the so-called 'Alten Reichsformat', a relatively flat brick of 25 x 12 x 6.5 cm, which emphasises the massiveness of the building (4). The integrated new archive building uses the same materials, adapted, however, to current technological developments in the form of double-skin masonry (5). The rear-ventilated construction consists of an inner concrete wall that envelops the archive, an insulation layer and ventilation layer as well as a masonry shell mounted on the outside with intermittent consoles to carry the loads. The bricks of the masonry shell are arranged at an offset to achieve a graphical texture and emphasise the monumentality of the building without creating monotony.

4

Ornaments in the masonry façade
The masonry façade uses ornaments to create texture across the large surface areas.

5

Transition from the existing to the new building
The windows of the existing building were filled in; the new building is windowless.

The façade of the new archive tower is a rear-ventilated solid façade. It consists of a 40 cm thick reinforced concrete wall, a ventilated insulation layer 20 cm thick and a 12 cm thick outer masonry shell. The loads of the outer shell are transferred with metal anchors which are anchored back in the loadbearing concrete wall. The graphic texture of the masonry is created by an offset of the individual bricks (6).

7

Exploded view of the façade construction
The support anchors that transfer the load of the outer masonry shell are clearly visible.

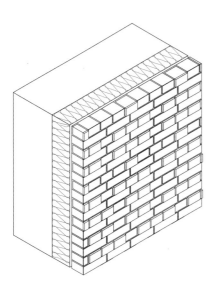

6

Façade construction
Isometric view of the façade construction consisting of loadbearing concrete wall, insulation, ventilation layer and outer masonry shell.

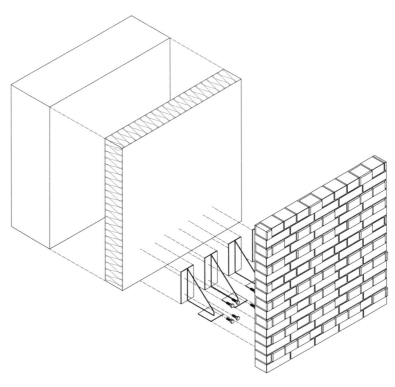

8

Detail of existing façade
The windows of the existing façade were filled in with masonry since the archive behind them does not require any daylight.

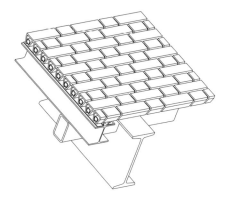

To underline the monumental character of the building the roof was also clad with bricks. The construction includes open joints to accommodate the drainage of rainwater. Therefore, the bricks were encased in metal brackets and mounted on a steel support structure. To emulate the texture of the façade the roof bricks were arranged in the same offset manner (9-11).

Since the bricks were laid with open joints rainwater must be drained inside the building through an internal drainage system. The resulting large air space underneath the roof also allowed the placement of necessary building services and facilities, protected from the elements and invisible from the outside.

9

Overview detail of the roof construction
The open roof construction also consists of bricks, mounted on a support structure.

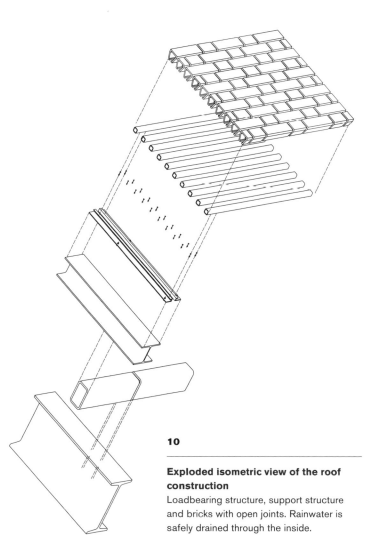

10

Exploded isometric view of the roof construction
Loadbearing structure, support structure and bricks with open joints. Rainwater is safely drained through the inside.

11

Connection masonry wall – roof
Connection between masonry wall and open roof with snow guard.

Post-and-beam façade

Project | New building for the Department for Architecture and Interior Design at the University of Applied Sciences
Location | Detmold, Germany
Completion | 2008
Client | Bau- und Liegenschaftsbetrieb NRW
Architect | werkstatt emilie
Structural engineering | Dr. Möller und Oberhokamp
Building services | Bau- und Liegenschaftsbetrieb NRW
Façade planning | werkstatt emilie
Façade contractor | Lanco GmbH

Conception

The design of the faculty building for the Department for Architecture and Interior Design at the University of Applied Sciences in Detmold, Germany (Hochschule Ostwestfalen-Lippe) was the result of a student competition; further planning was also done by a team of students. The approximately 97 m long and 23 m wide flat roof building is three storeys high. The work and lecture areas are arranged as open spaces in a free sequence, separated by so-called 'Prof-Boxes' in the shape of small cubicles. A central staircase provides access to the entire building which houses two lecture halls, administration and laboratory areas, a library and a cafeteria.

The objective of the design was to create an open structure that supports today's open teaching and learning methods on one hand, and allows for future modifications due to changes in the curriculum or working methods on the other. The result is a footprint of maximum flexibility, which serves the mandatory functions (fire protection, evacuation, sanitary facilities) and a few special functions but is mainly designed to be adaptable to an ever-changing educational institution and thus serves as a modifiable envelope. Therefore, the façade was developed as a non-hierarchic transparent building envelope with freely positioned apertures (1, 2).

1

Overall view
Faculty building for the Department for Architecture and Interior Design at the University of Applied Sciences in Detmold.

2

Façade view
Faculty building with its free and playful arrangement of opaque and transparent areas in various colours.

Façade

In order to reflect the flexible use of the interior in the outer appearance of the building, the post-and-beam construction was arranged in a free play of opaque and transparent surfaces. Approximately 70% of the surface area has storey-high glazing with exterior Venetian blinds (3). Only the northern side is not equipped with sun protection; interior glare protection and shades were, however, installed where needed. The opaque areas consist of fixed elements with a casement window for individual ventilation. Above these are mechanically driven ventilation flaps that ensure nighttime ventilation and cross ventilation to support the overall climate concept. Convector heaters to temper the building are integrated into the façade as well; they are also located at the closed surfaces below the opening flaps. In order to highlight the separation of façade and building structure, the ceiling projects across the structural grid, creating open areas particularly in the corner sections which are clad with almost transparent, stepped insulated glazing (4).

The post-and-beam system applied here is based on a structural grid of 1.35 m. It consists of a loadbearing structure made of laminated veneer timber, onto which an aluminium attachment system is mounted to accommodate the load of the glazing and panels. On the outside, the glass panes and panels are secured with aluminium clamping strips and a cover strip. Since the façade consists of storey-high glazing there is no need for horizontal beams in the transparent areas – therefore the interior glass panels are made of laminated safety glass for fall protection.

3

Façade layout
Storey-high glazing, casement windows and ventilation flaps are arranged in a free layout of opaque and transparent surfaces.

4

Corner area of the façade
View of fully glazed corners with recessed inner supports.

The overview detail (5, 6) shows the post-and-beam construction. It consists of storey-high vertical members which are horizontally connected with beams at the ceiling face in the lower as well as the upper area (7).

Ordinarily, post-and-beam constructions are not used to carry the building load, and are therefore structurally separated from the loadbearing construction. In this example the façade construction was also separated on each individual storey, and butted in the ceiling façade area. At the lower end the vertical members are fixed onto steel brackets. At the upper end, the member is secured with a steel bracket with vertical slot holes, allowing for the vertical members to move independently from the primary construction. Horizontal loads that bear on the façade, such as wind loads, are thus transferred from the façade into the loadbearing construction. The exterior sun protection, a Venetian blind, is also positioned in the ceiling area. The non-transparent areas are enclosed with sheet metal panels into which opaque casement windows and ventilation flaps are integrated (8, 9).

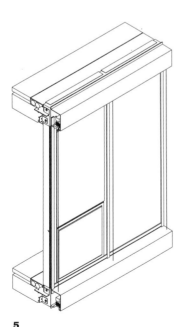

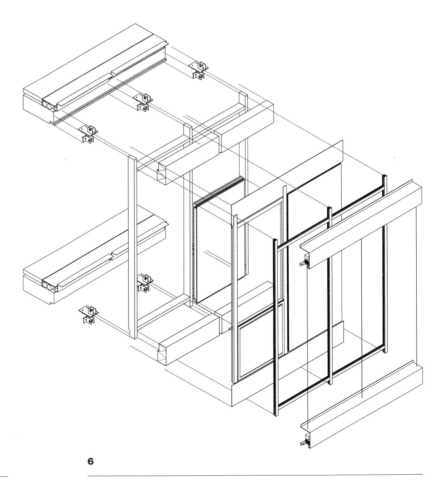

5

Isometric view of the post-and-beam façade
Façade consisting of storey-high vertical members, upper and lower horizontal members, and glazing.

6

Exploded isometric view of the post-and-beam façade
The individual constructional layers are clearly visible, consisting of inner vertical and horizontal members, the aluminium elements and the glazing.

The detail shows the arrangement of the post-and-beam construction with the laminated veneer timber located on the inside and thus protected from the elements. Mounted hereupon is the inner aluminium rail which accommodates the outer cover strips as well as the glass brackets. The distance between the rail and the outer profile provides for the necessary thermal separation. The outer aluminium clamping strip secures the glass against wind and impact forces, and seals off the glass panes with a continuous EPDM sealing profile. The butt end of the loadbearing timber construction is solved with a system node which allows for the invisible connection to carry the loads of the horizontal and the vertical members. Besides accommodating high glass loads as well, this engineered solution also offers easy assembly on site. In the opaque areas, the sheet metal panels and window elements are clamped into the post-and-beam construction in the same manner as the glass panes.

8

Nighttime view
The high transparency of the building is particularly visible at night; the opaque areas accentuate the overall appearance.

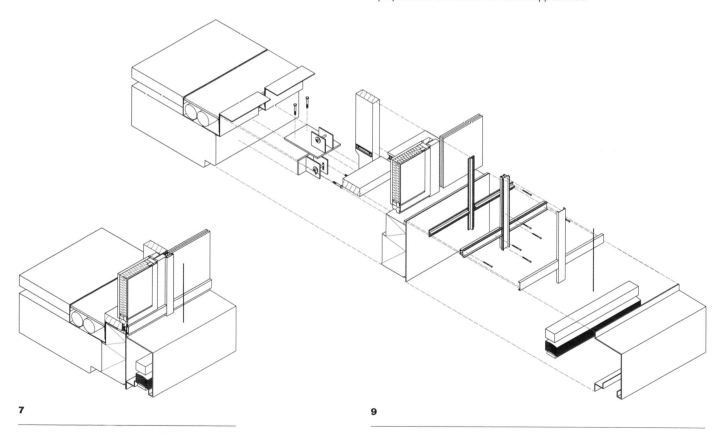

7

Isometric view of the post-and-beam detail
Post-and-beam system with vertical and horizontal laminated veneer timber members on the inside, the sealing profiles, the glazing and the aluminium cover strip on the outside.

9

Exploded isometric view of the detail
Geometric dissolution of the node with the individual constructional layers.

Unit system façade

Project | Headquarters Süddeutscher Verlag
Location | Munich, Germany
Completion | 2008
Client | SV-Hochhaus
Architect | GKK+Architekten – Prof. Swantje Kühn, Oliver Kühn
Structural engineering | WSP Deutschland AG
Building services | WSP CBP Technische Ausrüstung GmbH
Façade planning | R+R Fuchs Ingenieurbüro für Fassaden-technik

Conception

The image of a crystal was the starting point for the design of the new construction of the Süddeutscher Verlag. Accordingly, the fully glazed façade of the building was developed as a double façade with glass panes that are slightly tilted in opposite directions – the resulting textured reflection of the sky and the surroundings optically dissolve the 100 m high building. The building complex consists of a six-storey flat roof building, a connecting atrium and a relatively slender high-rise with a footprint of 24 x 24 m. The entire complex provides space for approximately 1,850 people (1, 2).

A lot of effort was put into the development of the building services concept in order to optimise primary energy consumption and operating costs. In addition to reducing energy consumption, the main focus lay on energy gain and energy storage in the building or using energy generated in the immediate vicinity (geothermal energy, seasonal storage of heat and cold in the ground). The double façade was an integral part of the conception; in addition to the ventilation function it is equipped with integrated building services installations and thus offers individual climate regulation for each office.

1

Overall view of the new headquarters Süddeutscher Verlag
The building complex consists of a high-rise and a six-storey flat roof building connected by an atrium.

Façade

Part of the façade concept is a division in façade elements that are slightly tilted at opposing angles. The fragmented reflections make the entire ensemble resemble a giant crystal. The concept was realised with a unit system façade with prefabricated and pre-glazed façade elements that were installed on site. The individual elements were dimensioned grid-wide and storey-high to allow for quick assembly by a small assembly team without scaffolding (3).

The double façade was designed as a box-window façade with an insulated glass unit on the inside and single glazing on the outside (4). Venetian blinds serving as sun protection are located in the wind-protected space between the glass panes. Operable windows on the inside provide access to this space for cleaning and maintenance (6).

The gap between the glass panes as well as the offices behind them are ventilated via slots in the butt area of the unit; the air from the offices is exhausted through the suspended ceiling in the hallways. Part of the building services concept is to utilise the façade to regulate the indoor climate by means of decentralised façade ventilation devices. Hereby, particular consideration was paid to individual control options. The system is designed as a 'hybrid ventilation' system; during the transitional seasons spring and autumn, natural ventilation occurs via the double façade, whereas mechanical ventilation is used during the other periods.

2

Façade of the high-rise building
The façade with glass panes tilted at opposing angles reflects the sky and the surroundings, creating a faceted and varied play of reflections.

3

Box-window façade
The unit system façade with its slightly tilted surfaces functions according to the principle of the box window. Decentralised façade ventilation units are integrated in the ceiling face.

The façade of the high-rise building is a unit system façade (9, 10). The frame sections consist of thermally separated aluminium sections which accommodate the different functions. In the ceiling area, conical sections – tapered toward the outside – cover the ventilation slot as well as the louvres of the sun protection.

The exploded isometric view (5) shows the outer single-pane glazing, the sun protection in between the inner and outer glazing and the aluminium frame that forms the unit. The casement windows visible on the inside are only used for cleaning purposes. The connectors in the upper area of the façade can also be seen; they are attached to the concrete slab of the building which is concealed by the double floor. The unit system façade elements are connected to each other with three sealing layers, whereby the inner layer controls vapour tightness and the two outer layers rain and wind tightness.

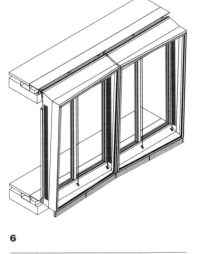

6

Isometric view of a façade unit
Two façade units with box-window areas as well as interior casement windows and flexible sun protection.

4

Front view façade
Even when seen frontally, the tilted glass panes and the resulting reflections are clearly visible.

5

Exploded isometric view of a façade unit
This isometric clearly shows the individual façade layers with insulating glass units on the inside, sun protection in the box-window gap and single glazing or flexible sun protection on the outside.

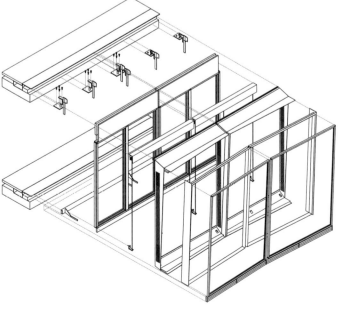

A detailed view of the butt area of the unit (7, 8) explains the concept behind the node, consisting of fall-proof single glazing as composite safety glass, held by the frame sections of the unit system façade. The frame is connected to adjacent units with the above mentioned three continuous plastic sealing profiles. The interior space is thermally separated by means of plastic spacers that carry the inner frame. The casement sash, also thermally separated, holds the inner insulating glazing. The space in the box window is thus outside of the thermal envelope and guards the sun protection against wind loads.

9

Interior of the foyer area
The area of the foyer shows the transparent character of the fully glazed façade particularly well. Here, the glass panes are also tilted and thus dissolve the austerity of the design.

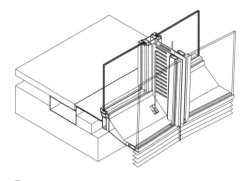

7

Exploded isometric view of the butt area of the element
Butt area of the element with exterior single glazing, the loadbearing aluminium section as well as the thermal separation layer and the inner window sash. The detail also clearly shows the three sealing layers between the elements.

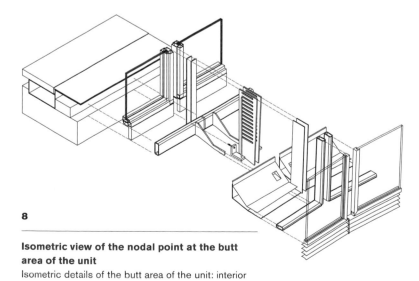

8

Isometric view of the nodal point at the butt area of the unit
Isometric details of the butt area of the unit: interior casement window, box-window space and single glazing on the outside.

10

View of the high-rise façade
Side view of the façade with different reflections of the sky.

8 | A Look Into the Future

Future façades will be subjected to influences from different directions: besides purely technical developments of traditional constructions made of steel, aluminium, glass, timber and mineral materials, the spectrum of new materials and their combination will become increasingly important. These materials – mostly integral, thus composed of different source materials – offer potential in terms of integrating new functions besides the usual ones of loadbearing and sealing. And energetic aspects will play a role: where does the energy come from, how can it be stored and how can we achieve lower consumption? Another topic for future considerations are systems: they become increasingly complex and serve more and more functions but they also begin to require specialists to install and operate them. When will we reach the end of this development? And ultimately there is the question of how we – as planners and designers – handle the new developments and which tools we will need. Exciting questions, for some of which answers are slowly emerging – however, merely as snapshots of our current state of knowledge and guaranteed to be subjected to change.

Undoubtedly, further development of façade materials will continue: the glass industry will achieve lower U-values for insulating glass by increasing the application of already available triple glazing, and aluminium and steel section producers will also improve their systems. Alternative solutions in the form of composite materials or timber will increase their market share in façade systems, driven by energetic, constructional or design objectives.

1

Aerated concrete block
A block incorporating areas of different density combines the functions of loadbearing and insulating.

In the field of mineral materials, the race for ever lower U-values will continue as well; in the medium term, this will likely lead to a linkage between the functions insulating and loadbearing. On one hand, the U-values of materials typically used for loadbearing functions such as bricks or concrete will come closer to those of pure insulating materials, achieved by new production technologies or a combination of different qualities (1, 2). On the other hand, insulating materials will be improved in terms of their potential as loadbearing elements, or will be further developed into holistic systems by adding additional components. Another aspect that has great impact on materials development is the life cycle discussion, driven by shortage of resources and the resulting necessity to recycle as well as the consequences in terms of material choice and construction methods.

2

Lightweight concrete
Expanded concrete is used to produce a lightweight concrete, which has good insulating properties in addition to its loadbearing capacity.

Material and construction

Developments in the area of membrane façades are promising in the realm of design as well as of construction and function. Besides the familiar pneumatic cushion structures, new solutions are developed that emulate double façades with a textile outer skin. They are not only interesting in terms of appearance and construction but also in their performance capacity. Initial approaches to complete systems already exist. It is self-evident that performative foils and membranes used as industrial materials as well as in the clothing industry are considered for use in the building envelope – particularly due to their low weight, principally easy exchangeability and the possibility of targeted and therefore economical use in construction (3, 4).

4

**Unilever House, Hamburg,
Behnisch Architekten, 2009**
The double façade consists of an inner aluminium façade and an outer asynclastically formed membrane façade.

3

Water Cube, Beijing, China, PTW Architects, 2008
Pneumatic building envelope consisting of an inner and outer ETFE cushion structure.

Two directions of development can be identified in the field of composite materials: on the one hand, large-sized building parts are employed more frequently due to improved manufacturing options and simpler transportation and assembly possibilities. Lamination technologies from boatbuilding and wind energy technology are adapted for this purpose. On the other hand, pultrusion sections – sections made of glass-fibre reinforced plastic – are a commendable material for façade systems due to their low thermal conductivity (5, 6).

As an alternative for plastic composites, i.e. crude oil based materials, biodegradable as well as renewable materials such as hemp or flax can be used as bio composites. However, until now, these materials have found very limited application even though first façade projects from bio composites have been realised. The problem is the resin because it is still typically based on crude oil, but the goal of current research is to develop resins from natural or renewable resources.

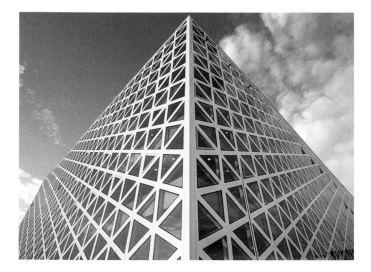

5

6

Hogeschool Windesheim, Building X, Zwolle, the Netherlands, Broekbakema Architecten, 2011
Composite façade of the administration building with glued fixed glazing.

Façade elements at Hogeschool Windesheim, Building X, Zwolle
The large elements of the composite façade reduce the effort of sealing the individual elements.

For glass structures the general goal remains to further increase the transparency of the construction. This is done either by dividing the loadbearing structure into more and more filigree structural elements or by transferring the entire loadbearing function to the glass itself, which becomes a structural material. Glass structures can, for example, become self-supporting by specific deformation; the shape of the glass pane increases planar rigidity and eliminates the need for additional support by means of bar-shaped loadbearing structures (7-10).

7

INHolland seminar building, Delft, the Netherlands, Rijk Rietveld Architects, 2011
Cable-mesh façade with composite parts that are partially integrated into the glass layer. Pre-tensioning the cables reduces a deflection of the extremely slender façade.

8

Apple Store, New York, Bohlin Cywinski Jackson, 2006
The glass cube is carried by 9 m high glass fins and a glass support system.

9

Casa da Música, Porto, Portugal, OMA, 2005
The curved shape of the glass panes not only gives the façade a softer
appearance but it also results in a structural rigidity of the glass layer that
eliminates the need for an additional support structure.

On the level of microscopic material development, the performance parameters of the materials take centre stage. Nano gel, an insulating material consisting mostly of encapsulated air, has been used as insulating filling material for several façade components. The same is true for Phase Change Materials (PCM), which can be employed as energy storage in surfaces close to the façade to store and release energy in a controlled manner. Polymer technology is another subject of several research projects related to the material's potential for property changes of building parts and components. A polymer coating can, for example, offer a switchable change of the surface – to become reflecting, absorbing or change in colour. Such adaptive surfaces offer great potential for the building sector; however, a lot more development work is needed to comply with current and future requirements and needs.

10

Vakko Headquarters, Istanbul, Turkey, REX Architecture, 2010
Storey-high glass panes are stiffened by X-shaped thermal deformation.

Climate, comfort, energy

Climatic evaluation of buildings and the integration of this knowledge into the design process mandate particular attention – even if knowledge and technology in this field have improved considerably and should be common standard. In terms of technological realisation different solutions in the building envelope can facilitate the integration of building services components (11, 12); such integrated façades were described in chapters 1 and 5 of this book. Following several façades that were essentially developed as one-offs, system solutions are now available that can be integrated into the design and building process as a finished product to accommodate some building services functions in addition to the actual façade function. However, the difficulty of integrating system solutions into the design process and adjusting and coordinating the individual planning and execution subcontractors has proven more of an innovation roadblock than any technical problems.

Nowadays, building services components are integrated into the façade even with refurbishment projects, particularly because separating building services from the building and integrating them into the façade means less interference with the existing building (13).

11

E² Façade concept
Product solution of a double façade with integrated building services components (Schüco).

12

NEXT Façade concept
Product conception of an open-system integrated façade (Alcoa).

13

Refurbishment Stadtsparkasse Ludwigshafen
Integrated façade as a refurbishment solution for an existing administrative building.

In addition to regulating the energy consumption of the building, energy gained by the building is another important factor. Independent of economic considerations, photovoltaic technology has improved considerably in terms of efficiency and scope of variation such as different colours, for example. Solutions are discussed that, in addition to a purely southerly oriented installation, focus less on maximum energy yield and more on balancing out solar peaks at noontime, thereby allowing for more variation in the geometric orientation. Exploiting photosynthesis in algae offers new outer appearances in addition to its energy gain potential, which is similar to that of simple PV modules. Following a project study, the first prototype was built with the BIQ House (Splitterwerk/Arup) as part of IBA Hamburg 2013 (14).

Considering the energy necessary for the production of materials and constructions – so-called grey or embedded energy – is a topic that will influence the choice of materials and constructions to a great extent. This energy is evaluated in terms of different factors that impact the environment. Consequently, the goal is to reduce the amount of material used as well as the possibility of deconstruction with the objective to reuse as many components and materials as possible – in addition to reducing the energy needed for production (15).

14

Project study algae façade, Hamburg, 2013
Research project of a planar biochemical reactor façade with energetic utilisation.

15

Aluminium waste
Aluminium waste from façades: the sections for thermal separation as well as coatings pose the biggest recycling problem.

Production and assembly

Analogous to the development of digital technologies the requirements of free-formed façades increase as well, causing the need for processing technologies for existing systems and alternative production methods (17). For the latter, new construction possibilities arise, made possible by the direct digital control of the production process, so-called 'file to factory' processes (16). Additive manufacturing methods is another promising topic in this context. Hereby, materials are joined in layers according to their specific properties to create building parts or components. The material portfolio, which originated from the field of thermoplastics, now encompasses composites and metals as well. Thus, this technology allows first steps toward fully functional building parts that can be applied to the façade.

17

Parametric concept
Concept for a free-form façade, based on a post-and-beam system (Schüco).

16

Façade plug connection 'Freessysteem'
Digitally controlled processing exemplified by a CNC milled plug connection (development: P. Stoutjesdijk).

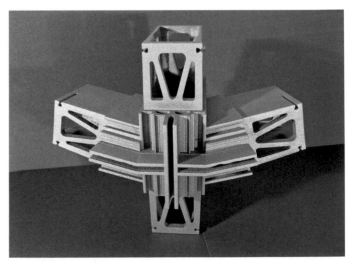

18

NEMATOX node
Façade node for free-form façades produced with additive aluminium print technology (Alcoa).

(Design) tools

Research has shown that for façade production, process sequence and process safety has increased significantly, allowing for short response times in case of changes. But it is also apparent that the communication between the disciplines participating in the planning and development of façades remains to be one of the greatest roadblocks in this process (19, 20). The general perception is still critical in spite of information and communication platforms. The thought of complete digital project communication, for instance based on BIM, does seem to make sense. However these systems must be carefully investigated in terms of their influence on individual interests with a potential limitation of the variety of products.

In parallel, tools to generate complex geometries and the integration of various parameters into the design process have developed at a rapid pace – they now offer simpler and more intuitive operation in addition to increasing complexity. As a result of such 'file to factory' systems complex constructions can be realised in terms of planning, construction and logistics.

19

Kunsthaus Graz, Peter Cook and Colin Fournier, 2003
Façade detail of the polycarbonate façade. The Kunsthaus was one of the early projects for which free geometry was translated into a buildable construction by means of a digital planning process.

20

Twin Tree Towers, Seoul, BCHO Architects, 2010
The geometrically free-form façade requires well prepared production and assembly logistics.

Authors

Professor Dr.-Ing. Ulrich Knaack was trained as an architect and was in practice in Düsseldorf. Today, he is Professor for Design of Construction and Building Technology at the Delft University of Technology, the Netherlands and Professor for Façade Technology at the TH Darmstadt in Germany. Author of several well-known German reference books on glass in architecture, co-author of the 'Imagine' book series and partner of the façade planning office imagine envelope in Rotterdam.

Professor Dipl.-Ing. Thomas Auer is a partner at Transsolar Energietechnik, Stuttgart, München, New York. He specialises in the field of integrated building services. Since 2001, Auer has been teaching at Yale University in the field Environmental Design of Buildings. In 2014 he became Professor for Building Technology and Building Climate at the Munich University of Technology.

Dr.-Ing. Tillmann Klein is an architect and heads the Façade Research Group at the Chair of Professor Knaack, Delft University of Technology. He is involved in teaching and research and is co-author of the 'Imagine' book series. He heads the façade planning office imagine envelope in Rotterdam.

Dr.-Ing. Marcel Bilow is part of the Façade Research Group at the Chair of Professor Knaack, Delft University of Technology and is involved in teaching and research at the chairs for Product Development and Architectural Engineering. He heads the prototype workshop Bucky Lab at Delft University of Technology, is co-author of the 'Imagine' book series and heads – together with Tillmann Klein – the façade planning office imagine envelope in Rotterdam.

Selected bibliography

History and general documentation

Francis D.K. Ching
Building Construction Illustrated
John Wiley, New York
3rd edition, 2000

Mike Davies
„A Wall for All Seasons",
in: RIBA Journal, 1981, vol. 88, no. 2.

Edward R. Ford
The Details of Modern Architecture
Cambridge, Mass., MIT Press, 1990

Thomas Herzog, Roland Krippner, Werner Lang,
Facade Construction Manual, Birkhäuser Verlag,
Basel and Edition Detail, Munich, 2004

Le Corbusier, Précisions sur un état présent
de l'architecture et de l'urbanisme
Editions Vincent, Fréal & Cie., Paris, 1929 –
English translation: Precisions: On the Present
State of Architecture and City Planning,
Cambridge, Mass., MIT Press, 1991

Christian Schittich (ed.)
Building Skins – Concepts, Layers, Materials
Birkhäuser Verlag, Basel and Edition Detail,
Munich, 2001

Technology

Andrea Compagno
Intelligent Glass Facades –
Material, Practice, Design
Birkhäuser, Basel, 5th edition, 2002

Klaus Daniels
Advanced Building Systems –
A Technical Guide for Architects and Engineers
Birkhäuser, Basel, 2003

Klaus Daniels, Dirk U. Hindrichs
Plusminus 20/40 Latitude – Sustainable Building
Design in Tropical and Subtropical Regions
Edition Axel Menges, Stuttgart, 2002

Johann Eisele, Ellen Kloft (eds.)
High-Rise Manual – Typology and Design,
Construction and Technology
Birkhäuser, Basel, 2003

Gerhard Hausladen, Michael de Saldanha,
Petra Liedl, Christina Sager
Climate Design – Solutions for Buildings
that Can Do More with Less Technology
Birkhäuser, Basel, 2005

Anette Hochberg, Jan-Henrik Hafke,
Joachim Raab
Open Close – Windows, Doors, Gates,
Loggias, Filters
Birkhäuser, Basel, 2010

Othmar Humm, Peter Toggweiler
Photovoltaics in Architecture
Birkhäuser, Basel, 1993

Patrick Loughran
Falling Glass – Problems and Solutions
in Contemporary Architecture
Birkhäuser, Basel, 2003

Eberhard Oesterle, Rolf-Dieter Lieb, Martin Lutz
Double-Skin Facades
Prestel, Munich, 2001

Just Renckens
Facades and Architecture –
Fascination in Aluminium and Glass
ed. by Federation of European Window and
Curtain Wall Manufacturers' Association,
Frankfurt, 1998

Materials

Manfred Hegger, Volker Auch-Schwelk,
Matthias Fuchs, Thorsten Rosenkranz
Construction Materials Manual
Birkhäuser, Basel
and Edition Detail, Munich, 2005

Patrick Loughran
Failed Stone – Problems and Solutions
with Concrete and Masonry
Birkhäuser, Basel, 2006

Axel Ritter
Smart Materials in Architecture,
Interior Architecture and Design
Birkhäuser, Basel, 2006

Christian Schittich, Gerald Staib,
Dieter Balkow, Matthias Schuler,
Werner Sobek,
Glass Construction Manual
Birkhäuser, Basel, 1999

Els Zijlstra
Material Skills – Evolution of Materials
Materia, Rotterdam, 2005

Future

Ulrich Knaack
'About Innovation', in:
Adaptive Architecture Conference
The Building Centre, London, 2011

Ulrich Knaack, Sharon Chung-Klatte,
Reinhard Hasselbach
Prefabricated Systems – Principles of
Construction
Birkhäuser, Basel, 2012

Ulrich Knaack, Tillmann Klein, Marcel Bilow,
Holger Techen
'Total Concrete', in:
Beton Bauteile: Entwerfen, planen, ausführen,
Edition DBZ Deutsche BauZeitschrift, 2013

Tillmann Klein
Integral Façade Construction –
Towards a New Product Architecture
for Curtain Walls
Dissertation, TU Delft, 2013

Ulrich Knaack, Tillmann Klein
The Future Envelope 3
IOS Publisher, TU Delft, 2011

Marcel Bilow
International Façades – CROFT:
Climate Related Optimized Façade Technologies
Dissertation, TU Delft, 2012

Ulrich Knaack, Marcel Bilow, Linda Hildebrand,
Thomas Auer
Imagine 05: Energy
Publisher 010, Rotterdam, 2011

Holger Strauss
AM Envelope – The Potential of Additive
Manufacturing for Façade Construction
Dissertation, TU Delft, 2013

Ulrich Knaack, Marcel Bilow, Holger Strauss
Imagine 04: Rapids
Publisher 010, Rotterdam, 2010

Index

Academy Mont Cenis, Herne 38, 99
Activated building components 78
Adaptive façade 80, 85, 87, 101
Additive manufacturing technologies 126
Aerogel 91
Air inlets and outlets 31, 58, 96
Algae façade 125
Alternating façade 8. 33, 100
Aluminium 10, 21, 45, 48-50, 53, 61, 67-69,
103, 104, 109, 111-115, 118, 125
Arab World Institute, Paris 13
Apple Store, New York 122
ARAG Tower, Düsseldorf 12, 95-96
Arup 125
Atlasgebouw, Wageningen 43
Auer + Weber + Partner 92, 98

Baer, Steve 90
Balloon framing 23
Banco Mineiro de Produção, Belo Horizonte 12
BCHO Architects 127
Behnisch Architekten 120
Bio composites 121
BIQ House 125
Bohlin Cywinski Jackson 122
Bothe Richter Teherani 99
Box window 12, 20, 30-32, 87-88, 94-96,
103-105, 115, 117
Box-window façade 30-31, 93-94, 115
Brick 106-109
Brick façade 106-109
Broekbakema Architecten 121

Cable-mesh façade 41, 122
Capricorn House, Düsseldorf 101
Casa da Música, Porto 123
Cathedral of Amiens, Amiens 17
Ceiling-floor unit 59, 69, 81
Central Library, Seattle 83
Chek Lap Kok Airport, Hong Kong 66
Chilled ceiling 79
CNC milling 126
Cold façade 14-15
Collector façade 89-90
Comfort level 11, 71-78, 101-102
Composite 118, 121, 126
Concept House, RDM Campus, Rotterdam
102
Concrete 37, 58, 62, 66, 78
Concrete façade 66
Cook, Peter 127
Corridor façade 31, 93, 96-97
Crown Hall, Illinois Institute of Technology,
Chicago 44
Curtain wall 13, 27, 28

Daimler-Chrysler Building,
Potsdamer Platz, Berlin 35, 94
Davies, Mike 35, 89
debis Headquarters, Potsdamer Platz,
Berlin 35
Debitel Headquarters, Stuttgart
8, 33, 60
Detached house, Corrales, New Mexico 90
Double façade 11, 29-38, 93-101, 141
Double glazing 21, 70, 83, 92-93
Double-XX Office Building, Hamburg 99

E2 Façade concept 124
Energy demand 70, 77, 85, 100
Energy generation 36, 85, 100
Energy saving 87
Environmental control unit 101
ETFE 120
Exhaust air 30, 74, 76, 88, 90
Exhaust-air façade 88, 92-93
Expansion joints 39-40

Façade heating 78
Fan coil unit 11
Farnsworth House, Plano, Illinois
10-11, 24
Federal Center, Chicago 27
Fire protection 44, 50, 98, 110
File to factory 126, 127
Fittings 48, 50, 75
Fondation Cartier, Paris 21
Foster and Partners 12, 66, 95-96
Fournier, Colin 127
Fuller, Richard Buckminster 89

Gatermann + Schossig 29, 101
GKK+Architekten 114
Glass-in-lead technique 19
Gothic style 18
Greenhouse effect 87
Grid 42-44, 54
Gropius, Walter 19
GRP – glass-reinforced plastic 50, 121
Guggenheim Museum, Bilbao 7

Half-timbered construction 22
Hänsch, Klaus 41
Headquarters Süddeutscher Verlag, Munich
114
Heat transmission 86
Hogeschool Windesheim, Zwolle 121
Hopkins, Michael 92
Hybrid façade 100

Infill elements 37, 41
INHolland seminar building, Delft 122
Insulated glazing 21
Integrated façade 34, 100-101

Jahn, Helmut 34, 69, 77
Jewish Museum, Berlin 7
Joints 40, 51, 62-66, 68, 74-75
Jourda & Perraudin 38, 99
Juscelino Kubitschek Complex, Belo Horizonte
13

Kollhoff, Hans 35, 94
Kühn, Oliver 113
Kühn, Swantje 113
Kunsthaus Graz 127

Lateral forces 8, 16, 26, 59, 60
Le Corbusier 19, 88
Lehmbruck, Manfred 41
Libeskind, Daniel 7
Library, Delft University of Technology, Delft 25
Light intensity 71, 84
Light-directing elements 84, 100-101
Lightweight concrete 119
Lintel 16, 18
Lloyd's Building, London 35, 92
LMN Architects 83
Load transfer 38-39, 41, 44, 104
Louvres 47. 81-84, 95, 97, 101,116

Masonry 14, 16, 18, 54, 57-58, 106
Mecanoo Architecten 25
Mies van der Rohe, Ludwig
10-11, 19, 24, 27, 44
Modular façade 34-35
Mur neutralisant 88

New building for the Department for Architecture
and Interior Design at the University of Applied
Sciences, Detmold 110-113
New Parliament Building, London 92-93
Next Façade concept 124
Niemeyer, Oscar 12-13
Night-time cooling 78, 100
Noise 73-74, 93
Nouvel, Jean 13, 21

OMA 83, 123
Ortner & Ortner Baukunst 106
Otto, Frei 89

Petzinka, Pink und Partner 31, 35, 97
Phase Change Materials – PCM 123
Photonics Centre, Berlin 32, 95
Piano, Renzo 35
Plastic profile 49-50, 53
Platform framing 23
Plinth 58, 62
Point fixings 41
Polyvalent wall 35, 89
Port Event Center, Düsseldorf 15

Illustration credits

Post-and-beam construction 45-46, 111-112
Post-and-beam façade
25, 45, 59, 68, 110-113
Post Tower, Bonn 34, 69, 77
Predicted Percentage of Dissatisfied – PPD 72
Primary structure 37-43
Prisma Building, Frankfurt 92, 98-99
PTW Architects 120
Pultrusion 121

Rapid Manufacturing 126
RDM Campus 102
REX Architecture 123
RFR 8
Rijk Rietveld Architects 122
RKW Architektur + Städtebau
8, 12, 33, 60, 95-96, 100
Rogers, Richard 35, 89, 92
Room temperature 71-72, 77, 85, 93

Sauerbruch Hutton Architects 32, 95
Schneider + Schumacher 28
Screens 80
Sealing system 45, 51
Second-skin façade 30, 93, 98-99
Secondary structure 37-40, 42-43, 45, 47
Shaft-box façade 32, 76, 93, 95-96
Sick Building Syndrom – SBS 73
Silicone 21, 53, 64-66
Silk-screen printing 83
Single glazing
10-12, 19, 21, 52, 92-93, 115
Solar chimney 8, 92
Solar energy 80, 85-87, 90-91
Solid façade 106
Solid wall 14, 16
Splitterwerk 125
Stack effect 32, 95
Stadtsparkasse Ludwigshafen 124
Stadttor Building, Düsseldorf 31, 35, 96
Steel profile 50
Sun protection 70, 74, 80-84
Suspended structure 38
State Archive Nordrhein-Westfalen, Duisburg
106-109

Textiles 80, 82
Thermal bridge 11, 37, 54-55, 62
Thermal insulation 14, 20, 29, 49, 59, 62, 70
Thermal insulation layer 14-15, 59, 61
Thermal radiation 80, 86-87
Timber-frame construction 23
Timber window 48, 53, 63, 103
Tolerances 37, 44, 49, 51-52, 59, 67-69
Translucent materials 19, 91
Transparent heat insulation – THI 91
Triangle Building, Cologne 29
Trombe wall 90-91
TU Delft 102
Twin Tree Towers, Seoul 127

Unit system façade
46, 51, 58, 60, 69, 100, 114-117
User comfort
71-74, 76, 78, 84-85, 88, 92

Vakko Headquarters, Istanbul 123
Van den Oever, Zaaijer & Partners Architecten
43
Ventilation 20, 29-34, 47, 74-76, 93-101
Ventilation opening 75, 93, 96, 98

Wansleben, Norbert 15
Warm façade 14
Water Cube, Beijing 120
Weatherproofing layer 57-62
werkstatt emilie 110
Westhafen Haus, Frankfurt 28
Wilhelm Lehmbruck Museum, Duisburg
41
Wind-driven rain
44, 52-53, 57, 60-61, 64, 66
Wind loads
25-26, 29, 36, 38-40, 55, 82, 112
Wooden shingles 64

Chapter 1
3 Holger Knauf

Chapter 2
43 Holger Knauf

Chapter 3
15 Raico Bautechnik GmbH
19 Metallbau Erhard Holz GmbH, Leopoldshöhe

Chapter 4
37, 42, 43 Ilja Sucker

Chapter 5
2 Drawing based on Recknagel
4 Based on Germany Industry Norm
DIN EN ISO 7730
17 Ilja Sucker

Chapter 6
7 Le Corbusier, © VG Bild Kunst, Bonn 2007
8 Mike Davies, Richard Rogers Partnership,
London
12 Steve Baer, Zomeworks
13 NASA
16 Lloyd's Redevelopment, London,
Gartner GmbH
18 Alexandra Liedgens
23 Holger Knauf
34 Rouven Holz

Chapter 7
State Archive Nordrhein-Westfalen
1, 2, 3, 4, 5, 7, 11 Ortner & Ortner Baukunst
Headquarters Süddeutscher Verlag
1, 10 Alberto Ferrero
2, 3, 4, 9 Claus Graubner

Chapter 8
5, 7 Holland Composites
11, 17 Schüco
12, 18 Alcoa
16 Pieter Stoutjesdijk

We are especially grateful to these image provid-
ers. All other illustrations were created specifically
for this book or were provided by the authors.
Every reasonable attempt has been made to iden-
tify owners of copyright. If unintentional mistakes
or omissions occurred; we sincerely apologise
and ask for a short notice. Such mistakes will be
corrected in the next edition of this publication.

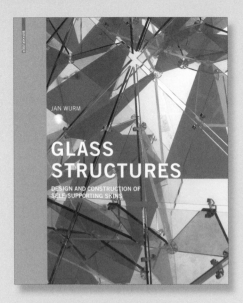

Glass Structures: Design and Construction of Self-supporting Skins
Jan Wurm

First comprehensive and systematic publication on glass as primary supporting framework

256 pages. 891 ills., 490 in colour.
23.0 x 29.7 cm. Hardcover.
ISBN 978-3-7643-7608-6 English

Scale: Support and Materialise. Columns, Walls, Floors
Alexander Reichel, Kerstin Schultz (eds.)

Design and construction of loadbearing structures and components

176 pages. 250 ills., 150 in colour.
22.0 x 28.0 cm. Softcover.
ISBN 978-3-0346-0040-8 English

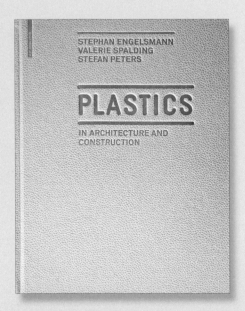

Plastics in Architecture and Construction
Stephan Engelsmann, Valerie Spalding, Stefan Peters

A systematic treatment of the use of plastics in architecture

176 pages. 242 ills., 118 in colour.
22.0 x 28.0 cm. Hardcover.
ISBN 978-3-0346-0322-5 English

Constructing Architecture Materials, Processes, Structures. A Handbook
Andrea Deplazes (ed.)

The prizewinning volume in its third, revised and expanded edition.

588 pages. 1740 ills.
24.0 x 29.7 cm. Softcover
ISBN 978-3-03821-452-6 English